Applied Soil Ecology

Applied Soil Ecology

Editor
Premjit Sharma

2014
Gene-Tech Books
New Delhi - 110 002

ISBN 81-89729-25-X

Published by : **GENE-TECH BOOKS**
4762-63/23, Ansari Road, Darya Ganj,
NEW DELHI - 110 002
Phone: 41562849
e-mail: genetechbooks@yahoo.co.in

Printed at : **Tarun Offset Printers**
Delhi

PRINTED IN INDIA

Preface

Soil ecology is the study of the interactions among soil organisms, and between biotic and abiotic aspects of the soil environment. It is particularly concerned with the cycling of nutrients, formation and stabilisation of the pore structure, the spread and vitality of pathogens, and the biodiversity of this rich biological community.

Soil is made up of a multitude of physical, chemical, and biological entities, with many interactions occurring among them. Soil is a variable mixture of broken and weathered minerals and decaying organic matter. Together with the proper amounts of air and water, it supplies, in part, sustenance for plants as well as mechanical support. The diversity and abundance of soil life exceeds that of any other ecosystem. Plant establishment, competitiveness, and growth is governed largely by the ecology below-ground, so understanding this system is an essential component of plant sciences and terrestrial ecology.

Modern research interests span many aspects of soil ecology and microbiology, Fundamentally, researchers are interested in understanding the interplay among microorganisms, fauna, and plants, the biogeochemical processes they carry out, and the physical environment in which their activities take place, and applying this knowledge to address environmental problems. Example research projects are to examine the biogeochemistry and microbial ecology of septic drain field soils used to treat domestic wastewater, the role of anecic earthworms in controlling the movement of water and nitrogen cycle in agricultural soils, and the assessment of soil quality in turf production.

This book emphasises the interrelations among plants, animals, and microbes, by establishing the fundamental physical and chemical properties of the soil habitat and then functionally characterizing the major components of the soil biota and some of their most important interactions. The fundamental principles underpinning soil ecology are established and this then enables an integrated approach to explore and understand the processes of soil nutrient cycling and the ecology of extreme soil conditions such as soil-water stress. This is book designed to meet the increasing challenges faced by today's environmental scientists, ecologists, agriculturalists, and biotechnologists.

Editor

Contents

1

Introduction

Soil is a major part of the natural environment. It is the result of the process of the gradual breakdown of rock—the solid geology that makes up the earth. As rock becomes broken down through a variety of processes, such as weathering and erosion, the particles become ground smaller and smaller. As a whole, soil is made up from four constituents: mineral material, organic material, air and water.

There are considered to be three main mineral parts to soil; 'sand', 'silt' and 'clay'. These parts give the soil its 'mineral texture'. In addition, as leaves and other organic material fall to the ground and decompose—there also forms an 'organic' layer. Soil scientists (or pedologists) use a series of sieves to separate out the constituent parts in order to characterise soil by texture class.

Soils may be characterised in terms of the properties they inherit from the underlying rock (the parent material) and the properties resulting from alteration of the original parent material by soil forming processes (also called 'pedogenic' processes). The latter effects can be observed in the surface and subsurface horizon layers. A good place to look at these layers is where a construction project has exposed a cross-section of earth, for instance where a new road is being built. The depth of soil varies geographically, but a depth of two metres in the UK is usually taken as the average depth to consolididated material (rock). Tropical soils can be tens of metres deep!

Formation of Soil

All soils initially come from rocks, this is termed the 'parent material'. The Parent Material may be directly below the soil, or great distances away if wind, water or glaciers have transported the soil. In addition to the soil parent material, soil formation is also dependent upon other prevailing processes affecting soil formation. The soil formation process is termed 'pedogenesis'. Climatic conditions are important factors affecting both the form and rate of physical and chemical weathering of the parent material. The formation of soils can be seen as a combination of the products of weathering, of structural development of the soil, of differentiation of that structure into horizons or layers, and lastly of its movement or translocation.

In fact there are many ways in which soil may be transported away from the location where it was first formed. In the UK, one of the most important of these factors is the effect that the ice ages have had. As glaciers pushed southwards down the country, they acted like giant bulldozers pushing soil ahead of them. Glacial 'till' or 'drift' deposits resulted many tens or even hundreds of miles from where the soils were first formed. When the weather warmed up and the glaciers melted, great mounds of soil were left behind, these are called 'moraines' and can be very impressive to see—clearly demonstrating the power of glaciers.

Water is also very important in moving soils. As rivers flow, so they transport soil particles along. If soil is washed into a river, the smallest particles will be carried the furthest by the water as they weigh the least. Heavier particles, such as sand, will be dropped earlier by the river. If you see an aerial photograph of a river as it flows into the sea or a lake, you may see the 'fan' of deposited soil materials. Soils deposited in this way around lakes and rivers are termed 'alluvial'. Soils deposited in lakes are called 'lacustrine', soils deposited by rivers 'riverine' and by sea 'marine' alluvial soils. Rainfall is also important in washing soil off exposed land. Even gentle rain can dislodge and move small soil particles. If this happens all across a field the process is termed 'sheet erosion'. Where there is a lot of rain, small gulleys can form which can speed up the process of 'gulley erosion'.

In tropical countries where there is extremely heavy rain, these gulleys can become so deep that you can walk down into them. Farmers should be aware of the effects of erosion upon their soil resource. Fields on slopes should be ploughed in lines across the gradient to prevent excess runoff and erosion. However, one can often see fields ploughed down the slope. This has the effect of creating small channels for rain water, and these channels can soon enlarge. One way scientists can try to quantify the effects of such erosion is to employ dye tracers in the field. To do this, during a rain event, a dye tablet is placed in the field and the extent of the resulting runoff pattern then becomes only too clear.

The wind is also able to move surprisingly large quantities of soil. On occasions fine soil deposits can be seen in many parts of the South of England which have been blown all the way from North African deserts. Soil in your garden is likely to be well mixed both as a result of years of digging and from the activities of animals in the soil such as worms. In agricultural fields there may be less mixing as there are fewer animals present due to the use of chemicals and compaction from heavy machinery. Ploughs are used to aerate and mix the soil instead. Where ploughs are continually used at a certain set depth, there may form a hard layer called a 'pan' below which there is little mixing. Such hard pans can also affect the drainage in the field where water cannot flow downwards affecting crop yields.

Weathering

One important influence on the formation of soils is weathering. There are two main categories of weathering, both having different effects:

Physical weathering

- *Freezing and Thawing*: Here, the expansive force of water pushes the soil structures apart. Water expands considerably when frozen and this expansion literally pushes the soil apart, breaking it down. When the ice thaws the soil can slump back again. The overall process is rather like a very slow 'churning'. Freeze-thawing can literally grind mountains down over time!
- *Heating and Cooling*: Here soils subjected to extremes of temperature are affected as they expand and contract. The effect is less pronounced that of freezing and thawing but over time this can become significant.
- *Wetting and Drying*: Soils that are wetted up may be prone to swelling. Clay minerals in particular exhibit this property. The soils that have thus

expanded then shrink when the soils dry out. These seasonal effects are termed shrinkage and swelling. Many household claims for subsidence are based upon such shrinkage and swelling of clays under foundations.

- *Grinding or Rubbing*: Most obvious on the beach, grinding of particles against each other leads to particle disintegration. This is why beach pebbles become smooth. Abrasion similarly breaks down the soil particles.
- *Organisms*: The effect of organisms, plants and animals, living in the soil cannot be overstated. Soil is home for a wide range of organisms. If plants can push through concrete—soil presents little obstacle! Worms churn their way through soil, mixing and aerating it all through their lives and there can be thousands of worms in a field.
- *Unloading*: When pressure is placed upon soil it becomes compressed. Never mind tractors, imagine the weight of a glacier! When ice melts a huge weight is lifted and the soils may react accordingly by uplifting and expanding.

Chemical weathering

- *Solution*: Certain solid components in the soil can be dissolved in soil water. In this way underground caverns can form in limestone karst landscapes. The name Karst comes from the Krass plateau in Slovenia where there are some of the most magnificent cave systems anywhere in the world—valley sized caves!
- *Hydrolysis*: Certain compounds in the soil can react to elements in the water.

- *Carbonation*: Soil compounds can react with carbonic acid.
- *Hydration*: Water in the soil can act to change the chemical structure of the soil components.
- *Oxidation*: Oxygen in the soil can act to change the chemical structure of the soil components.
- *Reduction*: A lack of oxygen in the soil can act to change the chemical structure of the soil components.

There are five key processes by which soil is formed, they are:

- *Leaching*: Leaching is the removal of soluable components of the soil column. As water washes down through the soil it can carry away bases such as calcium, held as exchangeable ions in clay-humus complexes, as well as acidification through the substitution of hydrogen ions.
- *Eluviation*: Here soil particles held in suspension, such as clay, are removed (*e.g.* washed away).
- *Illuviation*: Here soil particles held in suspension, such as clay, are accumulated (*e.g.* deposited).
- *Podsolisation*: Podsolisation occurs when strongly acid soil solutions cause the breakdown of clay minerals. As a result silica, aluminium and iron form complexes with organic substances in the soil. These minerals are removed from the surface zone of the soil and can accumulate in distinct dark subsurface layers—very evident on inspection. Upland heaths and moors often contain podsols.
- *Gleying*: Gleying occurs in waterlogged, anaerobic conditions when iron compounds are reduced and either removed from the soil, or segregated out as

mottles or concretions in the soil. Marshy wetlands often contain gleyed soils.

It is important to realise also that soil types are closely related to the shape of the landscape—or its 'topography'. Soil scientists use this to help them create soil maps. An experienced eye can determine changes in underlying soil types when walking through a landscape and observing changes in topography (and often vegetation too).

Structure of Soil

All the effects of soil forming processes, weathering, translocation leaching and so on leave the soil with a particular 'structure'. Soil structure refers to the shape, size and degree of development of the aggregation, if any, of the primary soil particles into naturally or artificially formed structural units and to their spatial arrangement. In describing soil structure, we refer to soil 'peds' and soil fragments or 'clods'. Soil peds are natural, relatively permanent aggregates, separated from each other by voids or natural surfaces of weakness.

Peds persist through cycles of wetting and drying. Soil Fragments and Clods are artificial structural units, formed at or near the surface by cultivation or frost action, and are not peds. Degradation of soil structures occur by a number of manners including:

- *Slaking*: The breaking down of aggregates by rain impact or under wet conditions; sorting soil particles and washing them into depressions—such slaking can form surface crusts or caps hindering shoot emergence in young crops. The winter of 2002 in the UK saw many fields in the South-East becoming 'capped' by the intense pounding of rain on the soil surface—which dramatically affected subsequent soil water runoff.

- *Cementation*: In certain cases, a subsoil iron 'pan' or hard ferrous layer can form, effectively preventing root penetration. Iron pans can be surprisingly hard, and if dug out can be held up and may need to be 'snapped' to break them!
- *Sodium leaching*: As saline soil solutions are diluted, sodium-saturated clays become unstable causing clay deflocculation and structural collapse.
- *Cultivation*: The downward compression from machinery when the soil is too 'plastic' can form a barrier to rooting and drainage as massive or platy structures form. This emphasises the need for the correct design of vehicles designed to work the soil. The following picture highlights a highly compressed soil, with a massive structure showing horizontal fracturing and stressed roots. So called plough pans' can also form if a plough of a set depth is used over and over again year after year. Where this happens a hard layer of soil can build up just under the depth of the plough. This hard layer can impede field drainage and adversely affect crop yield.

Soil Profile

The soil profile is one of the most important concepts in soil science. It is a key to understanding the processes that have taken in soil development and is the means of determining the types of soil that occur and is the basis for their classification. The soil profile is defined as a vertical section of the soil from the ground surface downwards to where the soil meets the underlying rock. The soil profile can be as little as 10 cm thick in immature soils and as deep as several metres in tropical areas where the climate is conducive to rapid alteration of the underlying rock to form soil.

In temperate areas, the soil profile is often around a metre deep and in arid areas somewhat shallower than this. Virtually all soil profiles are composed of a number of distinctive layers, termed horizons, interpretation of which is the key to understanding how the soil has formed. Most soils will have three or more horizons. Soils that have not been cultivated will normally have L, F and H layers at the surface. These layers largely represent different degrees of decomposition of organic matter, the L layer representing the litter layer formed of recognisable plant and soil animal remains, the F layer below, the fermentation layer, usually consisting of a mixture of organic matter in different stages of decomposition, and the H layer, the humose layer, consisting largely of humified material with little or no plant structure visible.

Below these, and in cultivated soils occupying the surface layer, is the A horizon composed of a more or less intimate mixture of mineral and organic matter. The A horizon is often referred to as the 'ploughed layer' in cultivated soils. It is an important part of the soil because it is a source of plant nutrients and contains the majority of plant roots. The A horizon may lie directly on the B horizon or, in well developed soils, there may be an intermediate leached horizon, termed E or A2, depending on the nomenclature system used. The E/A2 horizon is usually paler in colour than the horizons above and below because it is a horizon that has been subject to leaching and loss of components compared to the A and B horizons.

The B horizon is the horizon most widely used to identify soil types. Its morphology is important in supporting the classification of soils. In some soils the B horizon results purely from the weathering of the underlying rock whereas in other soils this weathering is supplemented by the translocation of materials from overlying horizons. Thus, the B horizon needs to be inspected carefully in order to understand the genesis of the soils.

B horizons may have a number of different subscripts indicative of the nature of the materials that have moved into the horizon, *e.g.* Bh indicates the translocation of humus into the horizon, Bs, the translocation of sesquioxides. These subscripts will vary according to the nature of the soil component that has accumulated but also with the system of nomenclature of soil types. Below the B horizon is the C horizon. This latter horizon is often consistent with the parent material and may have been little altered from the material in which the soil originally formed.

Most soils have A, B and C horizons. Some, generally weakly developed, soils may have A horizons lying directly on C horizons. When next you see a profile down through the soil, perhaps in an excavated pit or in a roadside cutting, take time to look at the profile and see if you can identify some of the different soil layers that make up the profile.

Types of Soil

Many natural bodies, such as plants and animals, are discrete entities which can be classified and guidelines for their identification followed. Soils are much more difficult to identify and classify than these discrete bodies for two main reasons: (i) soil is more or less a continuum covering the land surface of the earth, not a set of discrete entities; and (ii) most of the soil is below ground and therefore not readily visible. Soils grade into one another across the landscape usually without sharp boundaries between one type of soil and another.

Soil surveyors who make maps of soils have to use their skills in reading changes in the landscape coupled with auger borings in the soil to identify the nature of the soil. There are several ways of classifying a soil, from the simple to the complex. A soil type may be as simple as 'a sandy soil' or 'a clayey soil' and this is often the perception of many land users, such as farmers or civil engineers, who see it as material they have to deal with to achieve an end result, such as the growing of a crop of wheat, or the building a road.

Simple classifications tend to be of local and restricted relevance only. At the other end of the spectrum is the soil scientist who needs to understand how soils have formed, which types occur where, and for what the different types of soil can be used. The soil scientist seeks a much broader understanding, with the aim of underpinning the use and preservation of this important natural resource, and this has manifested itself in a number of detailed soil classification systems worldwide.

There is an enormous diversity of soils across the world. This is hardly surprising given the fact that soil formation and soil type are influenced by several key factors: the parent material, usually rock or sediment, but occasionally organic materials such as peat; climate, particularly temperature and rainfall; vegetation and other biota; topography/relief; time; and, increasingly, the influence of humans. All these factors will have an influence on soil development and hence soil type.

The potential for different combinations of these factors across the world is immense and hence it is not surprising that there are many thousands of different types of soil in the world, with different properties and potential. Just as Linnaeus produced a classification of plants that is in regular use today, over the centuries attempts have been made to develop classifications of soils on the basis of their properties, their genesis and in some cases their use potential. Since the late 19th century there have been several national classifications developed and used.

Most countries now have a system of nomenclature for their soil types. These classifications have grouped together similar soils and given names to the soils in the various classes. The classifications are mostly hierarchical in the sense that they consist of a number of levels, starting at the top with a number of major divisions, which can be subdivided into further subclasses and degrees of complexity. There have also been several major attempts at a world classification of soils, each of which have differed in the emphasis they have given to different soil properties. Such classification systems enable soil types to be classified and identified within a national and international framework.

The two most widely used international classifications are those of Soil Taxonomy (developed by the US Soil Conservation Service) and based on soil properties that can be objectively measured and observed and the FAO/ UNESCO legend/classification which is along broadly similar lines but not as precisely defined. It is regrettable that to date it has not been possible to produce an international classification that has been adopted worldwide but soil science is a young science with an important future and a unified classification system should eventually be developed. Until that is available, soil types should be named using both national nomenclature and have this linked to one or more of the international systems.

Soil Mapping

There are several thousand different types of soil throughout the world, a fact that is not surprising when bearing in mind the differences there are worldwide in the agents responsible for the building and forming of soil (landscape, climate, geology, vegetation, time and man). We know that there are at least this number of soil types because they have been mapped. In the past fifty years many countries of the world have been involved in making maps of their soils to determine the range of soil types in their territory, where the soils occur and how they can be used.

Soil mapping involves locating and identifying the different soils that occur, collecting information about their location, nature, properties and potential use, and recording this information on maps and in supporting documents to show the spatial distribution of every soil. Inorder to map and identify different types of soil it is necessary to have a system of soil classification. Just as in the case of plants where we have the Linnaean system, so an equivalent approach is needed for soils.

Soil classifications have provided a major challenge, one that has yet to be adequately solved. At present there is no single universally used system (unlike the case for plants). There are two international systems, FAO-UNESCO and US Soil Taxonomy, but a number of other systems have evolved and many countries have produced their own systems suited both to local conditions of soil formation and to local knowledge. However, the one uniform feature in all these systems is that they are based on the soil profile, that is the appearance of a section of the soil from the land surface through to the rock or other parent material that lies below the surface where the investigation is taking place. This examined thickness is usually about one metre but in tropical areas with deeper soil development it can be several metres, and in steep rocky landscapes may be much less than one metre.

Soils are allocated to a type according to their profile characteristics including, abundance of organic matter in the topsoil, the texture (combination of the different particle sizes and their ratio), colours representing aerobic or anaerobic conditions, structure of each layer, etc. Traditional soil mapping is conducted with an auger and spade at intervals throughout the landscape. The intervals between inspections can be according to a pre-determined grid (grid-survey) or, more often, are based upon the judgement of the surveyor who uses their knowledge of the inter-relationship between soil type and landscape, geology, vegetation, etc. to determine where to make inspections.

Auger borings are supplemented by excavated profile pits at determined points in the landscape. These profile pits are used to demonstrate lateral changes in the soil as well as vertical ones, and are important for the full description of type soils and for the taking of soil samples for chemical, physical and less commonly, biological laboratory analysis. In this way a picture is built up of the soil in a region and its relationship to the landscape in which it lies. Soils can be mapped at a range of scales from very detailed at 1:1,250 to 1:5,000 by which the pattern of soils in individual fields can be identified, through to scales of 1:500,000 to 1:500,000,000 which provide only a very generalised picture of the soils of a country or continent.

Sometimes soils are mapped with a specific aim in mind, such as the suitability of soils for a particular crop, suitability for irrigation, erosion risk and many other specific needs or environmental threats. Most organised soil surveys in the past have been general purpose surveys. These have the advantage that they provide the basis for many different uses, some of which may yet not be known.

Functions of Soil

Soils have many important functions. Perhaps the best appreciated is the function to support the growth of agricultural and horticultural crops. Soil is the mainstay of agriculture and horticulture, forming as it does the medium in which growth and ultimately the yield of food producing crops occurs. Farmers and gardeners have worked with their soils over many centuries to produce increasing amounts of food to keep pace with the needs of a burgeoning world population. The soil's natural cycles go a long way in ensuring that the soil can provide an adequate physical, chemical and biological medium for crop growth.

The farmer and horticulturalist have also become skilled in managing soils so that these natural cycles can be added to as necessary to facilitate adequate soil support and increasing yield to enhance production. Although it is difficult to rate the importance of the different soil functions, since all are vital to some extent, this function of supporting world agriculture and horticulture is a key one in the preservation and advancement of human life on this planet. As well as this agricultural and horticultural remit, soil also supports a wide range of trees grown for commercial purposes, *e.g.* plantations such as forests of commercially grown conifers.

Soil type is a fundamental factor in deciding what to grow and where to grow it—or rather optimising land use patterns. In addition to supporting world agriculture and horticulture, soil is also essential for maintaining natural and semi-natural vegetation, our forests, our grasslands and the huge breadth of species that occur worldwide. Generally in these situations, the reliance is on the soil's natural cycles to provide the necessary medium for this huge range of vegetation worldwide, in conjunction with the prevailing climate, landscape and the length of time the soils have been in existence. One of the best examples of this soil function is the tropical rain forest where more and more is now being learned of the important soil cycles that maintain this huge ecosystem. As well as being essential to agriculture, horticulture, forestry and natural and semi-natural systems, soil also plays an important role for our fauna.

The soil itself contains million of organisms, the exact nature and role of which we are still trying to determine. Undoubtedly the soil flora and fauna play a vital role in cycles which are fundamental to the ability of the soil to support natural and semi-natural vegetation without additions of fertiliser and other support mechanisms. They breakdown plant debris, take in components from the atmosphere, aerate the soil together with many other functions that make the soil such an important medium. It is quite staggering how much variety of life exists in soil.

One gramme of soil in a good condition can contain as many as 600 million bacteria belonging to up to 20,000 species! Even a similar amount of apparently barren desert soils can contain 1 million bacteria from up to 8,000 species. As well as those soil organisms that spend their whole life in the soil, there are also larger organisms that spend a part of their existence in soil but depend on the soil for important parts of their daily life, *e.g.* badgers, rabbits, prairie gophers, reptiles. Although it is not as well recognised as many of the other soil functions, the soil plays an important role for many birds. Some birds nest on its surface, many birds rely on it for a food supply and a very large number of birds rely on the vegetation for which the soil is responsible. Many birds also nest in burrows in the soil.

Soil is increasingly being recognised as playing a fundamental role in the quality and distribution of our water supply. The soil, coupled with the landscape and its vegetation is responsible for the distribution of all rainwater falling upon it. The nature of the topsoil will influence greatly whether the rainwater will run away across its surface, where it can supplement surface bodies of water, *e.g.* lakes, rivers, and in extreme situations lead to flash flooding, whether it will infiltrate to become stored in the soil for use by vegetation growing on it and by the soil based organisms, or whether it will flow through the soil to the reach the groundwater and at what rate it will do this.

The soil thus holds a key position with respect to our water supply cycle and is now seen as a key element as such by hydrologists. Related to how water moves through the soil and the absorption properties of soils is the soil's ability to perform an important function in pollution control. The fate and behaviour of pesticides in soils has become a key issue in relation to human health, in particular around the world. Soils differ greatly in their ability to manage applied pesticides.

Levels of organic matter in the topsoil, soil structure, particularly the presence of prismatic structure with large flow channels, and the texture and adsorption properties of the soils all affect the fate and behaviour (and 'attenuation') of pollutants in the soil. In recent decades the problem of managing nitrate levels in water has become a key issue as levels of nitrates in aquifers and surface water bodies continue to increase. Again the soil type and its inherent properties have become important in the attempts to regulate the movement downwards of the nitrate through the soil.

Soil has always been important for the foundation platform of buildings, roads and other communications and never more so. There is an ever increasing need to expand the number of dwellings and the supporting communication network. The problem with this is that buildings and engineering structures such as roads 'seal' the soil and negates the use of land for other purposes in the future. The nature of the soil, be it sandy, silty or clayey, has an important influence on the foundations to the structures and the measures that need to be taken to ensure stability to the structures.

Many people are interested in their origins and how earlier man lived. Soil plays an important part in the preservation of the earth's history. Archaeologists have come to recognise the importance of soils not only as a medium that has preserved evidence of the wares, properties and way of living of previous cultures and generations but also as a crucial component in determining the extent to which artefacts have been able to survive. Many village and town names today reflect the relationship between soil and society in the past—for instance 'Barton-le-Clay' in the UK.

Finally, soils have been recognised as having a key role in modifying and ameliorating the risks and effects of climate change. Soil organic matter is one of the major pools of carbon in the biosphere and is important both as a driver of climatic change and as a response variable to climate change, capable of acting both as a source and sink of carbon. There is significant interest in the fate of soil carbon, particularly the extent to which soils and land use may sequester carbon from the atmosphere, thus reducing CO_2 levels in the atmosphere or lose organic carbon to the atmosphere, thus increasing already existing levels in the atmosphere. Soils can also act as a source and sink for nitrous oxides and methane, both also important in creating the greenhouse gas effect.

Soils Preservation

Soils are one of the world's most precious commodities. It is probably safe to say that without soils there would be no support for life on this planet. The fact that the world's population is forecast to rise to from its current 5.8 billion to 8 billion by 2025, with a huge increase in demand for food highlights the reason why soils need to be preserved. For the past few decades there has often been a lack of appreciation of why we need to preserve soils. During this period there have been numerous examples of damage to world soils, some of which are now irreparable.

Some examples of the problems that face soils include:

Soil erosion and desertification: This is a major problem in many countries in the world, particularly, but not solely, in the drier, warmer parts of the world. Desertification currently affects about one-sixth of the world's population and a quarter of the world's land. Six to seven million acres are lost annually to soil erosion and both these processes continue to increase despite the many attempts made to introduce measures to stop or reduce erosion. It is a major threat to world soils and because of this there have been major initiatives to reduce desertification and erosion, including those by UNESCO, FAO, and the European and US Conservation bodies. However factors such as the need to increase food production and the likely impacts of global climate change are likely to make the fight against erosion and desertification increasingly difficult.

Salinisation and alkalisation: It is estimated that over 20 million hectares of land is affected by salinisation and alkalisation, again mainly the drier more arid areas of the world. Salts may be deposited in a number of ways; by winds coming in from the oceans, by use of irrigation with salt-containing water, and by salt accumulation in low-lying spots in the landscape as the climate varies. Like erosion and desertification, it is only in recent years that the problem has been recognised as a serious one and steps taken to reduce its spread. However, once accumulated in the soil it is a difficult and long-term problem to get rid of it.

Impact of intensive agriculture: The last four decades has seen a major increase in intensive agriculture in the bid to feed the world population more efficiently than ever before. In many countries, particularly the more developed countries, this intensification of agriculture has led to the use of more and heavier machinery, deforestation and clearing of land for use in cultivation. This has led to several problems including loss of organic matter, soil compaction and damage to soil physical properties generally, over-application of many nutrients in the form of fertilisers and pesticides, leading to problems with contamination of water supplies, potential loss of soil fauna and flora, such an essential part of a healthy soil, amongst others.

Urbanisation and land contamination: Urbanisation and the sealing of soils by infrastructures (*i.e.* building structures on top of soil) represents a significant loss of soil, particularly because the loss becomes more or less permanent. In developed countries there is a loss of about 1 per cent of land to urbanisation each decade.

The 18th and 19th centuries also saw enormous industrial expansion in many parts of the world with two major consequences for soils. Firstly, the widespread burning of fossil fuels in industry has led to widespread acidification of rainwater; for many decades soils of Western Europe have been leached by rainwater with a pH of about 4, *i.e.* strongly acidic. The second legacy of industry is contaminated land, the extent and effects of which are still not fully known. Thus in the last few decades there have been many pressures put on global soils and already significant and irrecoverable damages have been done.

We now need to move forward with much more awareness about the damage that has been caused to global soil resources and with a realistic programme for the preservation and protection of this unique natural entity, soil. Already there is increasing awareness of soil erosion and desertification in many countries and measures in place to protect soils from future damage. But in some environments, such as parts of Africa, the soil is a fragile entity, it lacks a robust nature, and therefore particular attention needs to be paid to its management and to ensuring that the management is clearly in tune with the soil. This is a real challenge but unless this challenge is accepted the route to future use of many of these soils could become a dead-end, with catastrophic results for many communities.

Many of the techniques needing to be applied are already well known, the challenge will be to integrate them into land use practice, particularly in a likely changing climate. The problems of salinisation and alkalisation continue and are reversible only with great difficulty. As with erosion and desertification, the causes are generally well established but very difficult and costly to remediate and solve. It is important that the problem does not continue to increase and there therefore needs to be better communication between the land user and those that advise them about their soils, and between policy makers and environmental scientists. The problems associated with intensification of agriculture outlined above are more readily treatable.

The causes of the problems are well established and already in many developed countries actions are being taken to work more in harmony with the soil. Some of the problems may take just a few years to improve upon whereas others, *e.g.*, low contents of organic matter and nitrates moving downwards into groundwater, may take decades to remediate. There need to be national and international plans for urbanisation that take into consideration the importance of protecting soils. Everyone is aware of the need to build more houses and to improve infrastructures but this process needs much more careful planning than before, with the need to preserve and protect soils in mind.

Already we are much more aware of the problems of soil contamination and the difficulties that arise from 'post-industrial' use contaminated land. Many countries now have safeguards in place to help to prevent soil contamination. Clean up of the contaminated land, however, remains a problem both technically and financially.

Soils have suffered from several decades of damage on a scale that we must not allow to continue. The reasons for this neglect are many. But if we are to ensure the protection of global soils into the future there must be much stronger awareness and acceptability among the public, policy makers and land users of the need to preserve soils for current and future generations. To ensure that this takes place it will be important that World Organisations and National Governments ensure there are adequate measures in place to give a much needed emphasis on soils and their future protection.

2

Description of Soil

A soil profile is exposed by a vertical cut through the soil. It is commonly conceived as a plane at right angles to the surface. In practice, a description of a soil profile includes soil properties that can be determined only by inspecting volumes of soil. A description of a pedon is commonly based on examination of a profile, and the properties of the pedon are projected from the properties of the profile. The width of a profile ranges from a few decimeters to several meters or more. It should be sufficient to include the largest structural units.

A soil horizon is a layer, approximately parallel to the surface of the soil, distinguishable from adjacent layers by a distinctive set of properties produced by the soil-forming processes. The term layer, rather than horizon, is used if all of the properties are believed to be inherited from the parent material or no judgment is made as to whether the layer is genetic.

The solum (*plural, sola*) of a soil consists of a set of horizons that are related through the same cycle of pedogenic processes. The solum of a soil presently at the surface, for example, includes all horizons now forming. It includes a bisequum. It does not include a buried soil or a layer unless it has acquired some of its properties by currently active soil-forming processes. The solum of a soil is not necessarily confined to the zone of major biological activity. Its genetic horizons may be expressed faintly to prominently. A solum does not have a maximum or a minimum thickness. Solum and soils are not synonymous. Some soils include layers that are not affected by soil formation. These layers are not part of the solum.

The number of genetic horizons ranges from one to many. An A horizon that is 10 cm thick overlying bedrock is by itself the solum. A soil that consists only of recently deposited alluvium or recently exposed soft sediment does not have a solum. Included are horizons with an accumulation of carbonates or more soluble salts if they are either within, or contiguous, to other genetic horizons and are judged to be at least partly produced in the same period of soil formation.

The lower limit, in a general sense, in many soils should be related to the depth of rooting to be expected for perennial plants assuming that water state and chemistry are not limiting. In some soils the lower limit of the solum can be set only arbitrarily and needs to be defined in relation to the particular soil. For example, horizons of carbonate accumulation are easily visualised as part of the solum in many soils in arid and semiarid environments. To conceive of hardened carbonate accumulations extending for 5 meters or more below the B horizon as part of the solum is more difficult.

Gleyed soil material begins in some soils a few centimeters below the surface and continues practically unchanged to a depth of many meters. Gleying immediately below the A horizon is likely to be related to the processes of soil formation in the modern soil. At great depth, gleying is likely to be relict or related to processes that are more geological than pedological. Much the same kind of problem exists in some deeply weathered soils in which the deepest material penetrated by roots is very similar to the weathered material at much greater depth. For some soils, digging deep enough to reveal all of the relationships between soils and plants is not practical.

Roots of plants, for example, may derive much of their moisture from fractured bedrock close to the surface. Descriptions should indicate the nature of the soil-rock contact and as much as can be determined about the upper part of the underlying rock. A sequum (plural, sequa) is a B horizon together with any overlying eluvial horizons. A single sequum is considered to be the product of a specific combination of soil-forming processes. Most soils have a single sequum, but some have two or more. A Spodosol, for example, can form in the upper part of an Alfisol, producing an eluviated zone and a spodic horizon underlain by another eluviated zone overlying an argillic horizon. Such a soil has two sequa.

Soils in which two sequa have formed, one above the other in the same deposit, are said to be bisequal. If two sequa formed in different deposits at different times, the soil is not bisequal. For example, a soil having an A-E-B horizon sequence may form in material that was deposited over another soil that already had an A-E-B horizon sequence. Each set of A-E-B horizons is a sequum but the combination is not a bisequum. The lower set is a buried soil. If the horizons of the upper sequum extend into the underlying sequum, the affected layer is considered part of the upper sequum.

For example, the A horizon of the lower soil may retain some of its original characteristics and also have some characteristics of the overlying soil. Here, too, the soils are not considered bisequal; the upper part of the lower soil is the parent material of the lower part of the currently forming soil. In many soils the distinction cannot be made with certainty. Nevertheless, the distinction is useful when it can be made. Where some of the C material of the upper sequum remains, the distinction is clear.

Pedons

Pedons representative of an extensive mappable area are generally more useful than pedons that represent the border of an area or a small inclusion. The description of a body of soil in the field, whether an entire pedon or a sample within it, should record the kinds of layers, their depth and thickness, and the properties of each layer. Generally, external features are observed throughout the extent of the polypedon; internal features are observed from the study of a pedon or that part of a pedon that is judged to be representative of the polypedon.

A pedon for detailed study of a soil is tentatively selected and then examined preliminarily to verify that it represents the desired segment of its range. A pit exposing a vertical face approximately 1 meter across to an appropriate depth is satisfactory for most soils. After the sides of the pit are cleaned of all loose material disturbed by digging, the exposed vertical faces are examined, usually starting at the top and working downward, to identify significant changes in properties. Boundaries between layers are marked on the face of the pit, and the layers are identified and described.

Photographs should be taken after the layers have been identified but before the vertical section is disturbed in the description-writing process. A point-count for estimation of the volume of stones or other features also is done before the layers are disturbed. A horizontal view of each layer is useful. This exposes structural units that otherwise may not be observable. Patterns of colour within structural units, variations of particle size from the outside to the inside of structural units, and the pattern in which roots penetrate structural units are often seen more clearly in a horizontal section.

Excavations associated with roads, railways, gravel pits, and other soil disturbances provide easy access for studying soils; old exposures, however, must be used cautiously. The soils dry out or freeze and thaw from both the surface and the sides. Frequently, the soil structure in such excavations is more pronounced than is typical; salts may accumulate near the edges of exposures or be removed by seepage; and other changes may have taken place.

Depth is measured from the soil surface. The soil surface is the top of the mineral soil; or, for soils with an O horizon, the soil surface is the top of the part of the O horizon that is at least slightly decomposed. Fresh leaf or needle fall that has not undergone observable decomposition is excluded from soil and may be described separately. The top of any surface horizon identified as an O horizon, whether Oi, Oe, or Oa, is considered the soil surface.

For soils with a cover of 80 percent or more rock fragments on the surface, the depth is measured from the surface of the rock fragments. The depth to a horizon or layer boundary commonly differs within short distances, even within a pedon. The part of the pedon that is typical or most common is described. In the soil description, the horizon or layer designation is listed and is followed by the values that represent the depths from the soil surface to the upper and lower boundaries, in that order.

The depth to the lower boundary of a horizon or layer is the depth to the upper boundary of the horizon or layer beneath it. The variation in the depths of the boundaries is recorded in the description of the horizon or layer. The depth limits of the deepest horizon or layer described include only that part actually seen. In some soils the variations in depths to boundaries are so complex that usual terms for description of topography of the boundary are inadequate. These variations are described separately.

For example, "depth to the lower boundary is mainly 30 to 40 cm, but tongues extend to depths of 60 to 80 cm." The lower boundary of horizon or layer and the upper boundary of the horizon or layer below share a common irregularity. The thickness of each horizon or layer is the vertical distance between the upper and lower boundaries. Thickness may vary within a pedon, and this variation should be shown in the description. A range in thickness may be given.

It cannot be calculated from the range of upper and lower boundaries but rather must be evaluated across the exposure at different lateral points. The location of upper and lower boundaries are commonly in different places. The upper boundary of a horizon, for example, may range in depth from 25 to 45 cm and the lower boundary from 50 to 75 cm. Taking the extremes of these two ranges, a wrong conclusion could be that the horizon ranges in thickness from as little as 5 cm to as much as 50 cm.

Land Surface Configuration

Land surface configuration considered here is geometrical and includes soil slope and land surface shape. Landform from a morphogenetic aspect is not considered. It may be applicable to a pedon or to a larger area.

Slope

Slope has a scale connotation. It refers to the ground surface configuration for scales that exceed about 10 meters and range upward to the landscape as a whole. Slope has gradient, complexity, length, and aspect. The scale of reference commonly exceeds that of the pedon and should be indicated. The scale may embrace a map unit delineation, component of it, or an arbitrary area. Slope gradient is the inclination of the surface of the soil from the horizontal. It is generally measured with a hand level.

The difference in elevation between two points is expressed as a percentage of the distance between those points. If the difference in elevation is 1 meter over a horizontal distance of 100 meters, slope gradient is 1 percent. A slope of 45° is a slope of 100 percent, because the difference in elevation between two points 100 meters apart horizontally is 100 meters on a 45° slope. Overland flow gradient is the slope of the soil surface in the direction of flow of surface water if it were present. The following examples show equivalences between percentage gradient and degree of slope angle:

Percentage	*Angle*	*Angle*	*Percentage*
00	°00'	0°	0
5	2°52'	2°	3.5
10	5°43'	4°	7.0
15	8°32'	6°	10.5
20	11°19'	8°	14.0
25	14°02'	10°	17.6
30	16°42'	12°	21.2
35	19°17'	15°	26.8

contd.

40	21°48'	20°	36.4
50	26°34'	25°	46.6
60	30°58'	30°	57.7
70	34°59'	35°	70.0
80	38°39'	40°	83.9
90	41°59'	45°	100.0
100	45°00'	50°	119.2

Slope Complexity refers to surface form on the scale of a mapping unit delineation. In many places internal soil properties are more closely related to the slope complexity than to the gradient. Slope complexity has an important influence on the amount and rate of runoff and on sedimentation associated with runoff.

Table 1. Definitions of slope classes

Classes		*Slope gradient limits*	
Simple Slopes	*Complex Slopes*	*Lower Percent*	*Upper Percent*
Nearly level	Nearly level	0	3
Gently sloping	Undulating	1	8
Strongly sloping	Rolling	4	16
Moderately steep	Hilly	10	30
Steep	Steep	20	60
Very steep	Very steep	>45	

Terms are provided for both simple and for complex slopes in some classes. Complex slopes are groups of slopes that have definite breaks in several different directions and in most cases markedly different slope gradients within the areas delineated. Significance of slope gradient is tied to other soil properties and to the purposes of soil surveys. Conventions are, therefore, provided in table 1 to adjust the slope limits of the various classes. Gently sloping or undulating soils, for example, can be defined to range as broadly as 1 to 8 percent or as narrowly as 3 to 5 percent.

Classes may exceed the broadest range indicated in table 1 by a percentage point or two where the range is narrow and by as much as 5 percent or more where the range is broad. If the detail of mapping requires slope classes that are more detailed than those in table 1, some of the classes can be divided as follows:

Nearly level	Level, Nearly level
Gently sloping	Very gently sloping, Gently sloping
Strongly sloping	Sloping, Strongly sloping, Moderately sloping
Undulating	Gently undulating, Undulating
Rolling	Rolling, Strongly rolling

In a highly detailed survey, for example, slope classes of 0 to 1 percent and 1 to 3 percent would be named "level" and "nearly level." Slope length has considerable control over runoff and potential accelerated water erosion. Terms such as "long" or "short" can be used to describe slope lengths that are typical of certain kinds of soils. These terms are usually relative within a physiographic region. A "long" slope in one place might be "short" in another. If such terms are used, they are defined locally.

For observations at a particular point, it may be useful to record the length of the slope that contributes water to the point in addition to the total length of the slope. The former is called point runoff slope length. The sediment transport slope length is the distance from the expected or observed initiation upslope of runoff to the highest local elevation where deposition of sediment would be expected to occur. This distance need not be the same as the point runoff slope length.

Slope aspect is the direction toward which the surface of the soil faces. The direction is expressed as an angle between 0 degree and 360 degrees or as a compass point such as east or north-northwest. Slope aspect may affect soil temperature, evapotranspiration, and winds received.

Surface Shapes of Land

Land surface shape has two components. One component is in a direction roughly parallel to the contours of the landform as seen from directly overhead. The other component of shape is a direction perpendicular to the contours; that is, the shape of the slope as seen from the side. The shape parallel to the contours is less commonly consistent for a soil than is the shape perpendicular to the contours.

The shape parallel to the contours can be described by the shape of the contours. The shape is linear if contours are substantially a straight line, as on the side of a lateral moraine. An alluvial fan has a convex contour, as does a spur of the upland projecting into a valley. A cove on a hillside or a cirque in glaciated landscapes has concave contours. In figure 1, the two upper blocks have concave contours and the two lower blocks have convex contours. Where the contour is convex, runoff water tends to spread laterally as it moves down the slope. Where the contour is concave, runoff water tends to be concentrated toward the middle of the landform.

The shape of the surface at right angles to the contours may also be described as linear, convex, or concave. Shape in this direction is usually identified simply as slope shape in contrast to slope contour in the other dimension. The surface of a linear slope is substantially a straight line when seen in profile at right angles to the contours. The gradient neither increases nor decreases significantly with distance. An example is the dip slope of a cuesta. On a concave slope (fig. 1), gradient decreases down the slope as on foot slopes.

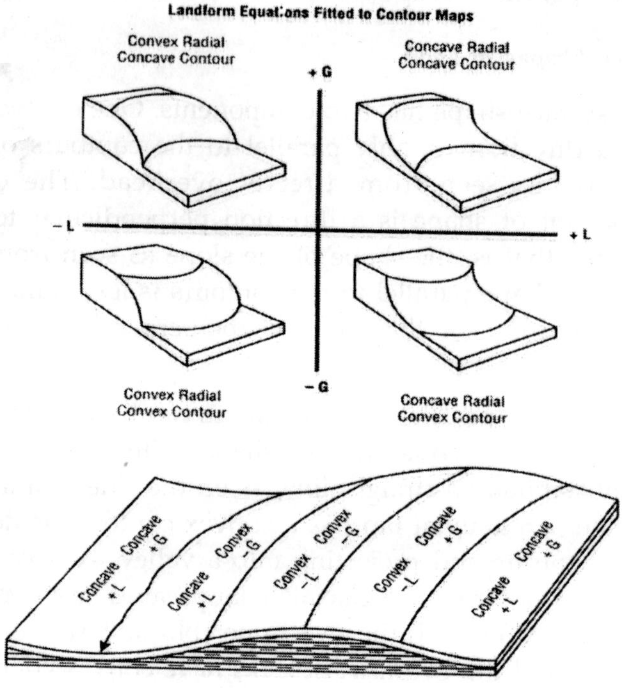

Figure 1. Illustrations of four combinations of concavity and complexity.

Runoff water tends to decelerate as it moves down the slope, and if it is loaded with sediment, the water tends to deposit the sediment on the lower parts of the slope. The soil on the lower part of the slope also tends to dispose of water less rapidly than the soil above it. On a convex slope (fig. 1), such as the shoulder or a ridge, gradient increases down the slope and runoff tends to accelerate as it flows down the slope. Soil on the lower part of the slope tends to dispose of water by runoff more rapidly than the soil above it. The soil on the lower part of a convex slope is subject to greater erosion than that on the higher part.

The configuration of the surface of a soil may be described in terms of both the shape of the contour and the shape of the slope. For example, a surface can be described as having a convex contour and a convex slope or a linear contour and concave slope. Description of an areal shape from the shape of two intersecting lines at right angles is applicable to all scales and does not require a contour map. The lines commonly would be parallel to and at right angles to the contour. Four line shapes are illustrated in figure 2: linear, hyperbolic concave, concave, and convex.

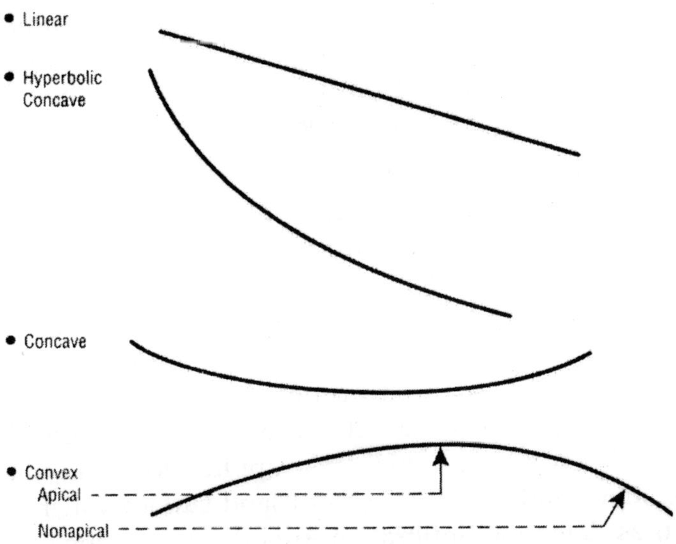

Figure 2. Four shapes of lines for description of land surface shapes.

Convex site may be usefully separated into apical and nonapical positions. Microrelief refers to differences in ground-surface height, measured over distances of meters. Naturally formed features contrast with those that are tillage-determined. In areas of similar relief, the surface may be nearly uniform, or it may be interrupted by mounds, swales, or pits. Examples include the microrelief created when trees are blown over, referred to as cradle-knoll microrelief. This consists of the knoll left by the earth that clung to the roots of the tree when it was uprooted and the depression from which it came.

Coppice dunes form where windblown soil material accumulates around widely spaced plants in arid regions. Gilgai produced by expansion and contraction of soils is a form of microrelief. Mima mounds and biscuit-scabland are other examples of microrelief, although individual mounds may cover 100 square meters or more. Descriptions should indicate whether mounds or depressions are closed, form a network, or are in a linear pattern. If mounds rest on a smooth surface, their size and spacing should be described.

At a specific site within an area having microrelief, it is important to note whether a described pedon is at a high point, on a slope, in a depression, or at some combination of these places. Internal soil properties in mounds may be different from the properties in depressions. Roughness refers to a ground surface configuration with a repeat distance between prominences of less than 50 cm and for areas less than about 10 m across. This scale applies to most tillage operations and affects aspects of land surface water flow such as detention, infiltration, runoff, and erosion.

Roughness pertains to the ground surface and includes rock fragments on the surface. It does not include vegetation. If vegetation is included, the fact should be indicated. Roughness along a line, referred to as one-dimensional roughness, can be measured more easily than can roughness for an area. Area measurements, however, permit the separation of random and tillage-determined roughness. The orientation to which the observation of one-dimensional roughness pertains must be specified relative to the direction of surface runoff or of air movement.

Position within the tillage-determined relief, if present, should be indicated for one-dimensional roughness. An example of such a position would be the non-traffic interrow in a tilled field. The standard deviation of the ground surface height is the primary descriptor. There are a number of approaches to the measurement of roughness, and those who are in agronomic disciplines should be consulted. The measurements depend on the variation in height from a levelled reference. Photographs may be used to illustrate the classes; placement in classes may be made directly from the photographs.

Vegetation and Soil

Correlations between vegetation and soils are made for three main purposes:

1. understanding soil genesis,
2. recognising soil boundaries, and
3. making predictions from soil maps about the kind and amount of vegetation produced.

The principal kinds of plants present are listed in order of their abundance. In annual cropland, the plant or plants that have been grown should be recorded, including significant weeds. In forested areas, separate treatment is often necessary for forest trees, understory of small trees and shrubs, and the ground cover. Many soils in range have an overstory of shrubs or low trees. These are listed separately from the grasses, forbs, and other ground cover. An idea of the density of stand or plant cover, such as average canopy cover of trees or shrubs, should be given.

The range in size of dominant species of trees can be given as "diameter breast height," if desired. Estimated percentage of the ground covered by grasses and forbs should be included. Common names of the plants may be used, if such names are clear and specific. In areas where the plants are important for the use and interpretation of the soil map, the soil survey record should include both common and scientific names of plants. If possible, the kinds and amounts of plants in the potential natural vegetation on a soil should be estimated. This vegetation is closely related to the soil and its genesis.

Generally, a close relationship exists between native vegetation and kinds of soil, yet there are important exceptions. Observations of the growth of native vegetation and cultivated crops aid in recognising soil boundaries and provide direct information about the behaviour of specific plants on different kinds of soil. Within fields of a single crop, differences of vigor, stand, or colour of the crop or of weeds commonly mark soil differences and are valuable clues to the location of soil boundaries.

By studying many sites of the same kind of soil under different land-use history, the potential plant community and principles of plant succession for that kind of soil can be ascertained, particularly if range and forestry specialists provide assistance. Farmers learn which crops do well and which do poorly on different kinds of soil and adjust their cropping patterns accordingly.

If the differences are large—as between crop failure and reasonable performance—the near absence of a given crop on a specific kind of soil questions the suitability of that kind of soil for the crop. If the differences are small, many non-soil factors can determine the farmer's choice of fields for a given crop. Yield information for cultivated crops, range, and trees should be associated with pedon descriptions insofar as possible.

Ground Surface

The ground surface of most soils is covered to some extent at least part of the year by vegetation. Furthermore, in many soils rock fragments form part of the mineral material at the soil surface. Together, the vegetal material that is not part of the surface horizon and the rock fragments form the ground surface cover. The proportion of cover, together with its characteristics, is very important in determining thermal properties and resistance to erosion. At one extreme, estimation of cover can be made visually without quantitative measurement. At the other extreme, transect techniques can be used to make a rather complete modal analyses of the ground surface.

More effort is justified on ground surface documentation if it is relatively permanent. In many instances, a combination of rapid visual estimates and transect techniques is appropriate. The ground surface may be divided into fine earth and material other than fine earth. The latter consists of rock fragments and both alive and dead vegetation. Vegetation is separated into canopy and noncanopy. A canopy component has a relatively large cross sectional area capable of intercepting rainfall compared to the area near enough to the ground surface to affect overland water flow.

In practice, the separation of canopy from noncanopy should be coordinated with the protocols for computation of susceptibility to erosion. Noncanopy material is commonly referred to as mulch. It includes rock fragments and vegetation. The first step in evaluation is to decide upon the ground surface cover components. The number is usually one to three. A common three-component land surface consists of trees, bushes, and areas between the two. The areal proportion of each component must be established. This may be done by transect.

If a canopy component is present, the area within the drip line as a percent of the ground surface is determined. For each canopy component, the effectiveness must be established. Effectiveness is the percent of vertical raindrops that would be intercepted. Usually the canopy effectiveness is estimated visually, but a spherical densitometer may be used. In addition to the canopy effectiveness, the mulch must be established for each component.

Transect techniques may be employed to determine the mulch percentage. The mulch can be subdivided into rock fragments and vegetation. From the areal proportions of the components and their respective canopy efficiencies and mulch percentages, the soil-loss ratio may be computed for the whole land surface. In addition to the observations for the computation of the soil-loss ratio, information may be obtained about the percent of kinds of plants, size of rock fragments, amount of green leaf area, and aspects of colour of the immediate surface that would affect absorption of radiant energy in an area.

Parent Material

Parent material refers to unconsolidated organic and

mineral materials in which soils form. The parent material of a genetic horizon cannot be observed in its original state; it must be inferred from the properties that the horizon has inherited and from other evidence. In some soils, the parent material has changed little, and what it was like can be deduced with confidence. In others, such as some very old soils of the tropics, the specific kind of parent material or its mode or origin is speculative. Much of the mineral matter in which soils form is derived in one way or another from hard rocks.

Glaciers may grind the rock into fragments and earthy material and deposit the mixture of particles as glacial till. On the other hand, rock may be weathered with great chemical and physical changes but not moved from its place of origin; this altered material is called "residuum from rock." In some cases, little is gained from attempting to differentiate between geologic weathering and soil formation because both are weathering processes. It may be possible to infer that a material was weathered before soil formation.

The weathering process causes some process constituents to be lost, some to be transformed, and others to be concentrated. Parent material may not necessarily be residuum from the bedrock that is directly below, and the material that developed into a modern soil may be unrelated to the underlying bedrock. Movement of soil material downslope is an important process and can be appreciable even on gentle slopes, especially on very old landscapes. Also, locally associated soils may form in sedimentary rock layers that are different.

Seldom is there certainty that a highly weathered material weathered in place. The term "residuum" is used when the properties of the soil indicate that it has been derived from rock like that which underlies it and when evidence is lacking that it has been modified by movement. A rock fragment distribution that decreases in amount with depth, especially over saprolite, indicates that soil material probably has been transported downslope.

Stone lines, especially if the stones have a different lithology than the underlying bedrock, provide evidence that the soil did not form entirely in residuum. In some soils, transported material overlies residuum and illuvial organic matter and clay are superimposed across the discontinuity between the contrasting materials. A certain degree of landscape stability is inferred for residual soils. A lesser degree is inferred for soils that developed in transported material.

Both consolidated and unconsolidated material beneath the solum that influence the genesis and behaviour of the soil are described in standard terms. Besides the observations themselves, the scientist records his judgment about the origin of the parent material from which the solum developed. The observations must be separated clearly from inferences. The lithologic composition, structure and consistence of the material directly beneath the solum are important.

Evidence of stratification of the material-textural differences, stone lines, and the like-need to be noted. Commonly, the upper layers of outwash deposits settled out of more slowly moving water and are finer in texture than the lower layers. Windblown material and volcanic ash are laid down at different rates in blankets of varying thickness. Examples of such complications are nearly endless. Where alluvium, loess, or ash are rapidly deposited on old soils, buried soils may be well preserved.

Elsewhere the accumulation is so slow that the solum thickens only gradually. In such places, the material beneath the solum was once near the surface but may now be buried below the zone of active change. Where hard rocks or other strongly contrasting materials lie near enough to the surface to affect the behaviour of the soil, their depths need to be measured accurately. The depth of soil over such nonconforming materials is an important criterion for distinguishing different kinds of soil. Geological materials need to be defined in accordance with the accepted standards and nomenclature of geology.

The accepted, authoritative names of the geological formations are recorded in soil descriptions where these can be identified with reasonable accuracy. As soil research progresses, an increasing number of correlations are being found between particular geological formations and the mineral and nutrient content of parent materials and soils. For example, certain terrace materials and deposits of volcanic ash that are different in age or source, but otherwise indistinguishable, vary widely in the content of cobalt.

Wide variations in the phosphorus content of two otherwise similar soils may reflect differences in the phosphorus content of two similar limestones that can be distinguished in the field only by specific fossils. Igneous rocks formed by the solidification of molten materials that originated within the earth. Examples of igneous rocks that weather to important soil material are granite, syenite, basalt, andesite, diabase, and rhyolite. Sedimentary rocks formed from sediments laid down in previous geological ages.

The principal broad groups of sedimentary rocks are limestone, sandstone, shale, and conglomerate. There are many varieties of these broad classes of sedimentary rocks; for example, chalk and marl are soft varieties of limestone. Many types are intermediate between the broad groups, such as calcareous sandstone and arenaceous limestone. Also included are deposits of diatomaceous earth, which formed, from the siliceous remains of primitive plants called diatoms.

Metamorphic rocks resulted from profound alteration of igneous and sedimentary rocks by heat and pressure. General classes of metamorphic rocks important as parent material are gneiss, schist, slate, marble, quartzite, and phyllite.

Material Produced by Weathering of Rock

The nature of the original rock affects the kinds of material produced by weathering. The rock may have undergone various changes, including changes in volume and loss of minerals-plagioclase feldspar and other minerals. Rock may lose mineral material without any change in volume or in the original rock structure, and saprolite is formed. Essentially, saprolite is a parent material. The point where rock weathering ends and soil formation begins is not always clear. The processes may be consecutive and even overlapping.

Quite different soils may form from similar or even identical rocks under different weathering conditions. Texture, colour, consistence, and other characteristics of the material should be included in the description of soils, as well as important features such as quartz dikes. Useful information about the mineralogical composition, consistence, and structure of the parent rock itself should be added to help in understanding the changes from parent rock to weathered material.

Transported Material

The most extensive group of parent materials is the group that has been moved from the place of origin and deposited elsewhere. The principal groups of transported materials are usually named according to the main agent responsible for their transport and deposition. In most places, sufficient evidence is available to make a clear determination; elsewhere, the precise origin is uncertain. In soil morphology and classification, it is exceedingly important that the characteristics of the material itself be observed and described.

It is not enough simply to identify the parent material. Any doubt of the correctness of the identification should be mentioned. For example, it is often impossible to be sure whether certain silty deposits are alluvium, loess, or residuum. Certain mud flows are indistinguishable from glacial till. Some sandy glacial till is nearly identical to sandy outwash. Fortunately, hard-to-make distinctions are not always of significance for soil behaviour predictions.

Waterblown Materials

Alluvium: Alluvium consists of sediment deposited by running water. It may occur on terraces well above present streams or in the normally flooded bottom land of existing streams. Remnants of very old stream terraces may be found in dissected country far from any present stream. Along many old established streams lie a whole series of alluvial deposits in terraces-young deposits in the immediate flood plain, up step by step to the very old deposits on the highest terraces. In some places recent alluvium covers older terraces.

Lacustrine deposits: These deposits consist of material that has settled out of bodies of still water. Deposits laid down in fresh-water lakes associated directly with glaciers are commonly included as are other lake deposits, including some of Pleistocene age that are not associated with the continental glaciers. Some lake basins in the Western United States are commonly called playas; the soils in these basins may be more or less salty, depending on climate and drainage.

Marine sediments: These sediments settled out of the sea and commonly were reworked by currents and tides. Later they were exposed either naturally or following the construction of dikes and drainage canals. They vary widely in composition. Some resemble lacustrine deposits.

Beach deposits: Beach deposits mark the present or former shorelines of the sea or lakes. These deposits are low ridges of sorted material and are commonly sandy, gravelly, cobbly, or stony. Deposits on the beaches of former glacial lakes are usually included with glacial drift.

Windblown Materials

Windblown material can be divided into groups based on particle size or on origin. Volcanic ash and cinders are examples of materials classed by both particle size and origin. Other windblown material that is mainly silty is called loess, and that which is primarily sand is called eolian sand. Eolian sand is commonly but not always in dunes. Nearly all textures intermediate between silty loess and sandy dune material can be found.

Volcanic ash, pumice, and cinders are sometimes regarded as unconsolidated igneous rock, but they have been moved from their place of origin. Most have been reworked by wind and, in places, by water. Ash is volcanic ejecta smaller than 2 mm. Ash smaller than 0.05

mm may be called "fine ash." Pumice and cinders are volcanic ejecta 2 mm or larger.

Loess deposits typically are very silty but may contain significant amounts of clay and very fine sand. Most loess deposits are pale brown to brown, although gray and red colours are common. The thick deposits are generally massive and have some gross vertical cracking. The walls of road cuts in thick loess stand nearly vertical for years. Other silty deposits that formed in other ways have some or all of these characteristics. Some windblown silt has been leached and strongly weathered so that it is acid and rich in clay. On the other hand, some young deposits of windblown material are mainly silt and very fine sand and are low in clay.

Sand dunes, particularly in warm, humid regions, characteristically consist of fine or medium sand that is high in quartz and low in clay-forming materials. Sand dunes may contain large amounts of calcium carbonate or gypsum, especially in deserts and semideserts.

During periods of drought and in deserts, local wind movements may mix and pile up soil material of different textures or even material that is very rich in clay. Piles of such material have been called "soil dunes" or "clay dunes." Rather than identify local accumulations of mixed material moved by the wind as "loess" or "dunes," however, it is better to refer to them as "wind-deposited material." Also important but not generally recognised as a distinctive deposit is dust, which is carried for long distances and deposited in small increments on a large part of the world. Dust can circle the earth in the upper atmosphere.

Dust particles are mostly clay and very fine silt and may be deposited dry or be in precipitation. The accumulated deposits are large in some places. An immense amount of dust has been distributed widely throughout the ages. The most likely sources at present are the drier regions of the world. Large amounts of dust may have been distributed worldwide during and immediately following the glacial periods.

Dust is an important factor affecting soils in some places. It is the apparent source of the unexpected fertility of some old, highly leached soils in the path of wind that blows from extensive deserts some hundreds of kilometers distant. It explains unexpected micronutrient distribution in some places. Besides dust, fixed nitrogen, sulphur, calcium, magnesium, sodium, potassium, and other elements from the atmosphere are deposited on the soil in varying amounts in solution in precipitation.

Glacial Drift

Several terms are used for material that has been moved and deposited by glacial processes. Glacial drift consists of all of the material picked up, mixed, disintegrated, transported, and deposited by glacial ice or by water from melting glaciers. In many places glacial drift is covered by a mantle of loess. Deep mantles of loess are usually easily recognised, but very thin mantles may be so altered by soil-building forces that they can scarcely be differentiated from the underlying modified drift.

Glacial till: This is that part of the glacial drift deposited directly by the ice with little or no transportation by water. It is generally an unstratified, heterogeneous mixture of clay, silt, sand, gravel, and sometimes boulders. Some of the mixture settled out as the ice melted with very little washing by water, and some was overridden by the glacier and is compacted and unsorted. Till may be found in ground moraines, terminal moraines, medial moraines, and lateral moraines.

In many places it is important to differentiate between the tills of the several glaciations. Commonly, the tills underlie one another and may be separated by other deposits or old, weathered surfaces. Many deposits of glacial till were later eroded by the wave action in glacial lakes. The upper part of such wave-cut till may have a high percentage of rock fragments. Glacial till ranges widely in texture, chemical composition, and the degree of weathering that followed its deposition. Much till is calcareous, but an important part is noncalcareous because no carbonate rocks contributed to the material or because subsequent leaching and chemical weathering have removed the carbonates.

Glaciofluvial deposits: These deposits are material produced by glaciers and carried, sorted, and deposited by water that originated mainly from melting glacial ice. Glacial outwash is a broad term for material swept out, sorted, and deposited beyond the glacial ice front by streams of melt water. Commonly, this outwash is in the form of plains, valley trains, or deltas in old glacial lakes. The valley trains of outwash may extend far beyond the farthest advance of the ice. Near moraines, poorly sorted glaciofluvial material may form kames, eskers, and crevasse fills.

Glacial beach deposits: These consist of rock fragments and sand. They mark the beach lines of former glacial lakes. Depending on the character of the original drift, beach deposits may be sandy, gravelly, cobbly, or stony.

Glaciolacustrine deposits: These deposits are derived from glaciers but were reworked and laid down in glacial lakes. They range from fine clay to sand. Many of them are stratified or varved. A varve consists of the deposition for a calendar year. The finer portion reflects slower deposition during the cold season and the coarser portion deposition during the warmer season when runoff is greater. Good examples of all of the glacial materials and forms described in the preceding paragraphs can be found.

In many places, however, it is not easy to distinguish definitely among the kinds of drift on the basis of mode of origin and landform. For example, pitted outwash plains can scarcely be distinguished from sandy till in terminal moraines. Distinguishing between wave-cut till and lacustrine material is often difficult. The names themselves connote only a little about the actual characteristics of the parent material.

Material Moved by Gravity

Colluvium is poorly sorted debris that has accumulated at the base of slopes, in depressions, or along small streams through gravity, soil creep, and local wash. It consists largely of material that has rolled, slid or fallen down the slope under the influence of gravity. Accumulations of rock fragments are called talus. The rock fragments in colluvium are usually angular, in contrast to the rounded, water-worn cobbles and stones in alluvium and glacial outwash.

Organic Material

Organic material accumulates in wet places where it is deposited more rapidly than it decomposes. These deposits are called peat. This peat in turn may become parent material for soils. The principal general kinds of peat, according to origin are:

- *Sedimentary peat*: the remains mostly of floating aquatic plants, such as algae, and the remains and fecal material of aquatic animals, including coprogenous earth.
- *Moss peat*: the remains of mosses, including Sphagnum.
- *Herbaceous peat*: the remains of sedges, reeds, cattails, and other herbaceous plants.
- *Woody peat*: the remains of trees, shrubs, and other woody plants.

Many deposits of organic material are mixtures of peat. Some organic soils formed in alternating layers of different kinds of peat. In places peat is mixed with deposits of mineral alluvium and/or volcanic ash. Some organic soils contain layers that are largely or entirely mineral material. In describing organic soils, the material is called peat if virtually all of the organic remains are sufficiently fresh and intact to permit identification of plant forms. It is called muck if virtually all of the material has undergone sufficient decomposition to limit recognition of the plant parts. It is called mucky peat if a significant part of the material can be recognised and a significant part cannot. Descriptions of organic material should include the origin and the botanical composition of the material to the extent that these can be reasonably inferred.

Contrasting Soil Materials

Changes with depth that are not primarily related to pedogenesis but rather to geological processes are contrasting soil materials if they are sufficient to affect use and management. The term discontinuity is applied to certain kinds of contrasting soil materials.

Unconsolidated contrasting soil material may differ in pore-size distribution, particle-size distribution, mineralogy, bulk density, or other properties. Some of the differences may not be readily observable in the field. Some deposits are clearly stratified, such as some lake sediments and glacial outwash, and the discontinuities may be sharply defined.

Contrasting materials can be confused with the effects of soil formation. Silt content may decrease regularly with depth in soils presumed to have formed in glacial till. The higher silt content in the upper part of these soils can be explained by factors other than soil formation. In some of these soils, small amounts of eolian material may have been deposited on the surface over the centuries and mixed by insects and rodents with the underlying glacial till. In others, the silt distribution reflects water sorting. Inferences about contrasting properties inherited from differing layers of geologic material may be noted when the soil is described.

Generally, each identifiable layer that differs clearly in properties from adjacent layers is recognised as a subhorizon. Whether it is recognised as a discontinuity or not depends on the degree of contrast with overlying and underlying layers and the thickness. For many soils the properties inherited from even sharply contrasting layers are not consistent from place to place and are described in general terms. The C layer of a soil in stratified lake sediments, for example, might be described as follows: "consists of layers of silt and clay, 1 to 20 cm thick; the aggregate thickness of layers of silt and that of the layers of clay are in a ratio of about 4 to 1; material is about 80 percent silt."

Erosion

Erosion is the detachment and movement of soil material.

The process may be natural or accelerated by human activity. Depending on the local landscape and weather conditions, erosion may be very slow or very rapid. Natural erosion has sculptured landforms on the uplands and built landforms on the lowlands. Its rate and distribution in time controls the age of land surfaces and many of the internal properties of soils on the surfaces.

The formation of Channel Scablands in the state of Washington is an example of extremely rapid natural, or geologic, erosion. The broad, nearly level interstream divides on the Coastal Plain of the Southeastern United States are examples of areas with very slow or no natural erosion. Landscapes and their soils are evaluated from the perspective of their natural erosional history. Buried soils, stone lines, deposits of wind-blown material, and other evidence that material has been moved and redeposited is helpful in understanding natural erosion history.

Thick weathered zones that developed under earlier climatic conditions may have been exposed to become the material in which new soils formed. In landscapes of the most recently glaciated areas, the consequences of natural erosion, or lack of it, are less obvious than where the surface and the landscape are of an early Pleistocene or even Tertiary age. Even on the landscapes of most recent glaciation, however, postglacial natural erosion may have redistributed soil materials on the local landscape.

Natural erosion is an important process that affects soil formation and, like man-induced erosion, may remove all or part of soils formed in the natural landscape. Accelerated erosion is largely the consequence of human activity. The primary causes are tillage, grazing, and cutting of timber. The rate of erosion can be increased by activities other than those of humans. Fire that destroys vegetation and triggers erosion has the same effect. The spectacular episodes of erosion, such as the soil blowing on the Great Plains of the Central United States in the 1930s, have not all been due to human habitation.

Frequent dust storms were recorded on the Great Plains before the region became a grain-producing area. "Natural" erosion is not easily distinguished from "accelerated" erosion on every soil. A distinction can be made by studying and understanding the sequence of sediments and surfaces on the local landscape, as well as by studying soil properties.

Landslip erosion

Landslip erosion refers to the mass movement of soil. Slides and flows are two kinds of landslip erosion. In the slide process, shear takes place along one or a limited number of surfaces. Slide movement may be categorised as slightly or highly deformed, depending on the extent of rearrangement from the original organisation. In flow movement the soil mass acts as a viscous fluid. Failure is not restricted to a surface or a small set of surfaces. Classes of landslip erosion are not provided.

Location of the mass movement relevant to landscape features generally and the size of the mass movement in terms of area parallel to the land surface and the depth may be indicated. Information about the time since the mass movement took place may be very useful.

Water erosion

Water erosion results from the removal of soil material by flowing water. A part of the process is the detachment of soil material by the impact of raindrops. The soil material is suspended in runoff water and carried away. Four kinds of accelerated water erosion are commonly recognised: sheet, rill, gully, and tunnel (piping).

- Sheet erosion is the more or less uniform removal of soil from an area without the development of conspicuous water channels. The channels are tiny or tortuous, exceedingly numerous, and unstable; they enlarge and straighten as the volume of runoff increases. Sheet erosion is less apparent, particularly in its early stages, than other types of erosion. It can be serious on soils that have a slope gradient of only 1 or 2 percent; however, it is generally more serious as slope gradient increases.
- Rill erosion is the removal of soil through the cutting of many small, but conspicuous, channels where runoff concentrates. Rill erosion is intermediate between sheet and gully erosion. The channels are shallow enough that they are easily obliterated by tillage; thus, after an eroded field has been cultivated, determining whether the soil losses resulted from sheet or rill erosion is generally impossible.
- Gully erosion is the consequence of water that cuts down into the soil along the line of flow. Gullies form in exposed natural drainage-ways, in plow furrows, in animal trails, in vehicle ruts, between rows of crop plants, and below broken man-made terraces. In contrast to rills, they cannot be obliterated by ordinary tillage. Deep gullies cannot be crossed with common types of farm equipment. Gullies and gully patterns vary widely. V-shaped gullies form in material that is equally or increasingly resistant to erosion with depth.

  U-shaped gullies form in material that is equally or decreasingly resistant to erosion with depth. As the substratum is washed away, the overlying material loses its support and falls into the gully to be washed away. Most-U-shaped gullies become modified toward a V shape once the channel stabilises and the banks start to spall and slump. The maximum depth to which gullies are cut is governed by resistant layers in the soil, by bedrock, or by the local base level. Many gullies develop headward; that is, they extend up the slope as the gully deepens in the lower part.

– Tunnel erosion may occur in soils with subsurface horizons or layers that are more subject to entrainment in moving free water than is the surface horizon or layer. The free water enters the soil through ponded infiltration into surface-connected macropores. Desiccation cracks and rodent burrows are examples of macropores that may initiate the process. The soil material entrained in the moving water moves downward within the soil and may move out of the soil completely if there is an outlet. The result is the formation of tunnels which enlarge and coalesce. The portion of the tunnel near the inlet may enlarge disproportionately to form a funnel-shaped feature often referred to as a "jug." Hence, the term "piping" and "jugging." The phenomenon is favoured by the presence of appreciable exchangeable sodium.

Deposition of sediment carried by water is likely anywhere that the velocity of running water is reduced-at the mouth of gullies, at the base of slopes, along stream banks, on alluvial plains, in reservoirs, and at the mouth of streams. Rapidly moving water, when slowed, drops stones, then cobbles, pebbles, sand, and finally silt and clay. Sediment transport slope length has been defined as the distance from the highest point on the slope where runoff may start to where the sediment in the runoff would be deposited.

Wind erosion

Wind Erosion in regions of low rainfall, can be widespread, especially during periods of drought. Unlike water erosion, wind erosion is generally not related to slope gradient. The hazard of wind erosion is increased by removing or reducing the vegetation. When winds are strong, coarser particles are rolled or swept along on or near the soil surface, kicking finer particles into the air.

The particles are deposited in places sheltered from the wind. When wind erosion is severe, the sand particles may drift back and forth locally with changes in wind direction while the silt and clay are carried away. Small areas from which the surface layer has blown away may be associated with areas of deposition in such an intricate pattern that the two cannot be identified separately on soil maps.

Degree of Erosion

The degree to which accelerated erosion has modified the soil may be estimated during soil examinations. The conditions of eroded soil are based on a comparison of the suitability for use and the management needs of the eroded soil with those of the uneroded soil. The eroded soil is identified and classified on the basis of the properties of the soil that remains. An estimate of the soil lost is described. Eroded soils are defined so that the boundaries on the soil maps separate soil areas of unlike use suitabilities and unlike management needs.

The depth to a reference horizon or soil characteristic of the soil under a use that has minimised accelerated erosion are compared to the same properties under uses that have favoured accelerated erosion. For example, a soil that supports native grass or large trees with no evidence of cultivation would be the basis for comparison of the same or similar soil that has been cleared and cultivated for a relatively long time. The depth to reference layers is measured from the top of the mineral soil because organic horizons at the surface of mineral soils are destroyed by cultivation. The depths to a reference layer must be interpreted in terms of recent soil use or history.

Cultivation may cause differences in thickness of layers. The upper parts of many forested soils have roots that make up as much as one-half of the soil volume. When these roots decay, the soil settles. Rock fragment removal can also lower the surface. The thickness of surficial zones that have been bulked by tillage should be adjusted downward to what they would be if water had compacted them. The thickness of a plowed layer of a specific soil cannot be used as a standard for either losses or additions of material because, as a soil erodes, the plow cuts progressively deeper.

Nor can the thickness of the uncultivated and uneroded A horizon be used as a standard for all cultivated soil, unless the A horizon is much thicker than the plow layer. If the horizon immediately below the plowed layer of an uneroded soil is distinctly higher in clay than the A horizon, the plow layer becomes progressively more clayey under continued cultivation as erosion progresses; the texture of the plow layer may then be a criterion of erosion. Comparisons must be made on comparable slopes. Near the upper limit of the range of slope gradient for a soil, horizons may normally be thinner than near the lower limit of the range for the same soil.

Roadsides, cemeteries, fence rows, and similar uncultivated areas that are a small part of the landscape as a whole or are subject to unusual cultural histories must be used cautiously for setting standards, because the reference standards for surface-layer thickness are generally set too high. In naturally treeless areas or in areas cleared of trees, dust may collect in fence rows, along roadsides, and in other small uncultivated areas that are covered with grass or other stabilising plants. The dust thus accumulated may cause the surface horizon to become several centimeters thicker in a short time.

For soils having clearly defined horizons, differences due to erosion can be accurately determined by comparison of the undisturbed or uncultivated norms within the limitations discussed. The physical conditions of the material in the plowed layer, the appearance and amount of rock fragments on the surface, the number and shape of gullies, and similar evidence are relied on. For many soils having almost no horizon expression, attempting to estimate the degree of erosion serves little useful purpose.

Accelerated Erosion

The classes of accelerated erosion that follow apply to both water and wind erosion. They are not applicable to landslip or tunnel erosion. The classes pertain to the proportion of upper horizons that have been removed. These horizons may range widely in thickness; therefore, the absolute amount of erosion is not specified.

Class 1: This class consists of soils that have lost some, but on the average less than 25 percent, of the original A and/or E horizons or of the uppermost 20 cm if the original A and/or E horizons were less than 20 cm thick. Throughout most of the area, the thickness of the surface layer is within the normal range of variability of the uneroded soil. Scattered small areas amounting to less than 20 percent of the area may be modified appreciably. Evidence for class 1 erosion includes

- a few rills,
- an accumulation of sediment at the base of slopes or in depressions,
- scattered small areas where the plow layer contains material from below, and
- evidence of the formation of widely spaced, deep rills or shallow gullies without consistently measurable reduction in thickness or other change in properties between the rills or gullies.

Class 2: This class consists of soils that have lost, on the average, 25 to 75 percent of the original A and/or E horizons or of the uppermost 20 cm if the original A and/or E horizons were less than 20 cm thick. Throughout most cultivated areas of class 2 erosion, the surface layer consists of a mixture of the original A and/or E horizons and material from below. Some areas may have intricate patterns, ranging from uneroded small areas to severely eroded small areas. Where the original A and/or E horizons were very thick, little or no mixing of underlying material may have taken place.

Class 3: This class consists of soils that have lost, on the average, 75 percent or more of the original A and/or E horizons or of the uppermost 20 cm if the original A and/or E horizons were less than 20 cm thick. In most areas of class 3 erosion, material below the original A and/or E horizons is exposed at the surface in cultivated areas; the plow layer consists entirely or largely of this material. Even where the original A and/or E horizons were very thick, at least some mixing with underlying material generally took place.

Class 4: This class consists of soils that have lost all of the original A and/or E horizons or the uppermost 20 cm if the original A and/or E horizons were less than 20 cm thick. In addition, Class 4 includes some or all of the deeper horizons throughout most of the area. The original soil can be identified only in small areas. Some areas may be smooth, but most have an intricate pattern of gullies.

3

Characteristics of Soil Ecology

Some features of good soil are:

- feels soft and crumbles easily,
- drains well and warms up quickly in the spring,
- does not crust after planting,
- soaks up heavy rains with little runoff,
- stores moisture for drought periods,
- has few clods and no hardpan,
- resists erosion and nutrient loss,
- supports high populations of soil organisms,
- has a rich, earthy smell,
- does not require increasing inputs for high yields,
- produces healthy, high-quality crops.

All these criteria indicate a soil that functions effectively today and will continue to produce crops long into the future. These characteristics can be created through management practices that optimise the processes found in native soils.

Soils are made up of four basic components: minerals, air, water, and organic matter. In most soils, minerals represent around 45% of the total volume, water and air about 25% each, and organic matter from 2% to 5%. The mineral portion consists of three distinct particle sizes classified as sand, silt, or clay. Sand is the largest particle that can be considered soil.

Sand is largely the mineral quartz, though other minerals are also present. Quartz contains no plant nutrients, and sand cannot hold nutrients—they leach out easily with rainfall. Silt particles are much smaller than sand, but like sand, silt is mostly quartz. The smallest of all the soil particles is clay. Clays are quite different from sand or silt, and most types of clay contain appreciable amounts of plant nutrients. Clay has a large surface area resulting from the plate-like shape of the individual particles. Sandy soils are less productive than silts, while soils containing clay are the most productive and use fertilisers most effectively.

Soil texture refers to the relative proportions of sand, silt, and clay. A loam soil contains these three types of soil particles in roughly equal proportions. A sandy loam is a mixture containing a larger amount of sand and a smaller amount of clay, while a clay loam contains a larger amount of clay and a smaller amount of sand.

Another soil characteristic—soil structure—is distinct from soil texture. Structure refers to the clumping together or "aggregation" of sand, silt, and clay particles into larger secondary clusters. If you grab a handful of soil, good structure is apparent when the soil crumbles easily in your hand. This is an indication that the sand, silt, and clay particles are aggregated into granules or crumbs. Both texture and structure determine pore space for air and water circulation, erosion resistance, looseness, ease of tillage, and root penetration. While texture is related to the minerals in the soil and does not change with agricultural activities, structure can be improved or destroyed readily by choice and timing of farm practices.

Soil Organisms

An acre of living topsoil contains approximately 900 pounds of earthworms, 2,400 pounds of fungi, 1,500 pounds of bacteria, 133 pounds of protozoa, 890 pounds of arthropods and algae, and even small mammals in some cases. Therefore, the soil can be viewed as a living community rather than an inert body. Soil organic matter also contains dead organisms, plant matter, and other organic materials in various phases of decomposition. Humus, the dark-coloured organic material in the final stages of decomposition, is relatively stable.

Both organic matter and humus serve as reservoirs of plant nutrients; they also help to build soil structure and provide other benefits. The type of healthy living soil required to support humans now and far into the future will be balanced in nutrients and high in humus, with a broad diversity of soil organisms. It will produce healthy plants with minimal weed, disease, and insect pressure.

Considering the natural landscape, you might wonder how native prairies and forests function in the absence of tillage and fertilisers. These soils are tilled by soil organisms, not by machinery. They are fertilised too, but the fertility is used again and again and never leaves the site. Native soils are covered with a layer of plant litter and/or growing plants throughout the year. Beneath the surface litter, a rich complexity of soil organisms decompose plant residue and dead roots, then release their stored nutrients slowly over time.

In fact, topsoil is the most biologically diverse part of the earth. Soil-dwelling organisms release bound-up minerals, converting them into plant-available forms that are then taken up by the plants growing on the site. The organisms recycle nutrients again and again with the death and decay of each new generation of plants. There are many different types of creatures that live on or in the topsoil. Each has a role to play. These organisms will work for the farmer's benefit if we simply manage for their survival. Consequently, we may refer to them as soil livestock. While a great variety of organisms contribute to soil fertility, earthworms, arthropods, and the various microorganisms merit particular attention.

Roots

Roots absorb the water and nutrients that are needed by the plants for photosynthesis and respiration. Roots in the soil play an important role in the activity of organisms. Roots are often a little leaky, and the material that they leak is referred to as root exudates. The area immediately around the root is known as the rhizosphere. The rhizosphere environment has a lower pH, and the soil atmosphere has lower O_2 and higher CO_2 concentrations. The rhizosphere is higher in soil organism activity due to increased food supply leaked by the root for the organisms to use. This includes: amino acids, organic acids, carbohydrates, nucleic acids, growth factors, enzymes, and soughed-off tissue.

Benefits for the plant of having a rhizosphere include enhanced N mineralisation, enhanced N_2 fixation, and nutrient solubilisation. The rhizosphere is defined as an intense zone of stimulated microbial activity around the root. Within the rhizosphere microbial numbers are much greater than in the bulk soil. Microbes in the rhizosphere can be arbitrarily subdivided into the following groups:

- Pathogenic (invades and kills plants)
- Beneficial (often symbiotic with plants)

- Harmful (normally non-pathogenic opportunists on plants)
- Saprophytic (live on dead plants)
- Neutral (no effect on plants)

The microbes listed above are all competing for the some resources (space, nutrients and carbon) in the rhizosphere. The rhizosphere is a battlefield between pathogenic and non-pathogenic microorganisms. While we are mostly concerned with deleterious and pathogenic bacteria in the rhizosphere, there are some micro-organisms present in the rhizosphere which are good for roots. An example of this is the actinomycete Streptomyces which secretes antibiotics and toxins into the soil which then inhibits the growth of other rhizosphere microorganisms. *e.g.* It prevents the spread of the pathogenic "damping-off" fungus.

Earthworms

Earthworm burrows enhance water infiltration and soil aeration. Fields that are "tilled" by earthworm tunneling can absorb water at a rate four to ten times that of fields lacking worm tunnels. This reduces water runoff, recharges groundwater, and helps store more soil water for dry spells. Vertical earthworm burrows pipe air deeper into the soil, stimulating microbial nutrient cycling at those deeper levels. When earthworms are present in high numbers, the tillage provided by their burrows can replace some expensive tillage work done by machinery. Worms eat dead plant material left on top of the soil and redistribute the organic matter and nutrients throughout the topsoil layer.

Nutrient-rich organic compounds line their tunnels, which may remain in place for years if not disturbed. During droughts these tunnels allow for deep plant root penetration into subsoil regions of higher moisture content. In addition to organic matter, worms also consume soil and soil microbes. The soil clusters they expel from their digestive tracts are known as worm casts or castings. These range from the size of a mustard seed to that of a sorghum seed, depending on the size of the worm. The soluble nutrient content of worm casts is considerably higher than that of the original soil (Table 1).

A good population of earthworms can process 20,000 pounds of topsoil per year—with turnover rates as high as 200 tons per acre having been reported in some exceptional cases. Earthworms also secrete a plant growth stimulant. Reported increases in plant growth following earthworm activity may be partially attributed to this substance, not just to improved soil quality.

Table 1. Selected nutrient analyses of worm casts compared to those of the surrounding soil.

Nutrient	Worm casts (Lbs/ac)	Soil (Lbs/ac)
Carbon	171,000	78,500
Nitrogen	10,720	7,000
Phosphorus	280	40
Potassium	900	140

Earthworms thrive where there is no tillage. Generally, the less tillage the better, and the shallower the tillage the better. Worm numbers can be reduced by as much as 90% by deep and frequent tillage. Tillage reduces earthworm populations by drying the soil, burying the plant residue they feed on, and making the soil more likely to freeze. Tillage also destroys vertical worm burrows and can kill and cut up the worms themselves. Worms are dormant in the hot part of the summer and in the cold of winter.

Young worms emerge in spring and fall-they are most active just when farmers are likely to be tilling the soil.

As a rule, earthworm numbers can be increased by reducing or eliminating tillage (especially fall tillage), not using a moldboard plow, reducing residue particle size (using a straw chopper on the combine), adding animal manure, and growing green manure crops. It is beneficial to leave as much surface residue as possible year-round. Cropping systems that typically have the most earthworms are (in descending order) perennial cool-season grass grazed rotationally, warm-season perennial grass grazed rotationally, and annual croplands using no-till.

Ridge-till and strip tillage will generally have more earthworms than clean tillage involving plowing and disking. Cool season grass rotationally grazed is highest because it provides an undisturbed (no-tillage) environment plus abundant organic matter from the grass roots and fallen grass litter. Generally speaking, worms want their food on top, and they want to be left alone. Earthworms prefer a near-neutral soil pH, moist soil conditions, and plenty of plant residue on the soil surface. They are sensitive to certain pesticides and some incorporated fertilisers.

Carbamate insecticides, including Furadan, Sevin, and Temik, are harmful to earthworms. Some insecticides in the organophosphate family are mildly toxic to earthworms, while synthetic pyrethroids are harmless to them. Most herbicides have little effect on worms except for the triazines, such as Atrazine, which are moderately toxic. Also, anhydrous ammonia kills earthworms in the injection zone because it dries the soil and temporarily increases the pH there. High rates of ammonium-based fertilisers are also harmful.

Arthropods

In addition to earthworms, there are many other species of soil organisms that can be seen by the naked eye. Among them are sowbugs, millipedes, centipedes, slugs, snails, and springtails. These are the primary decomposers. Their role is to eat and shred the large particles of plant and animal residues. Some bury residue, bringing it into contact with other soil organisms that further decompose it.

Some members of this group prey on smaller soil organisms. The springtails are small insects that eat mostly fungi. Their waste is rich in plant nutrients released after other fungi and bacteria decompose it. Also of interest are dung beetles, which play a valuable role in recycling manure and reducing livestock intestinal parasites and flies.

Nematodes

Another worm-like organism is the nematode. Nematodes are microscopic worms that feed on organic matter and other soil animals or infect plant roots. Under a 10x hand lens, nematodes appear as transparent, thread-like worms. Parasitic nematodes are the most important from an agricultural standpoint. Many plants are affected, such as tomatoes, carrots, potatoes, peas, alfalfa, turfgrass, and fruit trees. Nematodes can parasitise virtually all crops and ornamental plants and can cause significant economic damage by reducing both yield and quality. Properly taken samples from small field units can reduce production costs by allowing the grower to eliminate nematodes

Lance nematodes, Hoplolaimus spp., are large nematodes which are highly resistant to effects of temperature extremes and dry soil conditions. One species, H. columbus, causes severe damage to soybeans and cotton. Another more widely distributed species, H. galeatus, is primarily a pathogen on grasses. Lance nematodes feed externally along root surfaces but may also feed with at least part of the body embedded in the root. Larvae look similar to adults except that they are smaller. This group of nematodes is easily detected with soil sampling.

The life cycle of this nematode takes about thirty days under ideal conditions, and females lay eggs singly rather than in a mass. Though a female may lay as many as one hundred eggs in a lifetime, that lifetime may last an entire growing season. Root-knot nematodes, Meloidogyne spp., are one of the important plant-parasitic nematodes because of their wide host range and widespread distribution. Root-knot larvae enter roots of host plants near root tips and remain inside the root at one location throughout their life. As larvae feed, the root cells divide rapidly near the nematode's head. This rapid cell division and enlargement cause the swelling or knots on roots..

Fungi

Fungi are not plants. Fungi were listed in the Plant Kingdom for many years. Then scientists learned that fungi show a closer relation to animals, but are unique and separate life forms. Now, Fungi are placed in their own Kingdom. The part of the fungus that we see is only the "fruit" of the organism. The living body of the fungus is a mycelium made out of a web of tiny filaments called hyphae. The mycelium is usually hidden in the soil, in wood, or another food source. A mycelium may fill a single ant, or cover many acres. The branching hyphae can add over a half mile (1 km) of total length to the mycelium each day. These webs live unseen until they develop mushrooms, puffballs, truffles, brackets, cups, "birds nests," "corals" or other fruiting bodies.

If the mycelium produces microscopic fruiting bodies, people may never notice the fungus. The most active decomposers of organic materials in a forested soil are the soil fungi. This is mainly because they are tolerant of acid soil conditions. All of us have seen fungi. Their size varies from single-cell yeasts to molds and mushrooms. The woody residue of the forest floor provides an abundance of food for certain fungi that are effective decomposers of lignin.

Most fungi build their cell walls out of chitin. This is the same material as the hard outer shells of insects and other arthropods. Plants do not make chitin. Fungi feed by absorbing nutrients from the organic material in which they live. Fungi do not have stomachs. They must digest their food before it can pass through the cell wall into the hyphae. Hyphae secrete acids and enzymes that break the surrounding organic material down into simple molecules they can easily absorb. Fungi have evolved to use a lot of different items for food. Some are decomposers living on dead organic material like leaves.

Some fungi cause diseases by using living organisms for food. These fungi infect plants, animals and even other fungi. Athlete's foot and ringworm are two fungal diseases in humans. The mycorrhizal fungi live as partners with plants. They provide mineral nutrients to the plant in exchange for carbohydrates or other chemicals fungi cannot manufacture. You probably use fungal products every day without being aware of it. People eat mushrooms of all shapes, sizes and colours. Yeasts are used in making bread, wine, beer and solvents. Drugs made from fungi cure diseases and stop the rejection of transplanted hearts and other organs.

Mycorrhizae are fungi associated with the fine roots of most plants. The term itself means "fungus root". There are hundreds species of fungi which function as mycorrhiza; most are basidiomycetes, the class of fungi which form mushrooms. An individual plant may have several different mycorrhizae associated with its roots, and some mycorrhizae may be limited to only a few species of plant. These fungi can benefit plants by enhancing the nutrient absorbing ability of roots.

Mycorrhizae are especially important in facilitating uptake of phosphorous. This enhancement of nutrient uptake is a result of the extensive system of hyphae and mycelia (thread-like filaments of the mycorrhizal fungus) that pervade soils. They function like root hairs but are much more far reaching.

The relationship of this fungus with plants is a mutually beneficial one, with the fungi receiving energy in the form of carbohydrates from the host plant. There are two types of mycorrhizae. Filaments of the first type, called ectomycorrhizae, penetrate between cells of roots, but not into root cells, and also form a thick cylindrical sheath around young lateral roots. The affected roots become short and thickened, and are deficient in root hairs. It is believed that this sheath may protect roots from invasion by plant pathogens. Most trees have this type of mycorrhizae.

A second type, called endomycorrhizae, actually penetrate into root cells and extend out from roots like root hairs to absorb nutrients from the soil solution. They do not form a sheath around roots nor do they alter the structure of roots. Endomycorrhizae are found on a greater range of plants than are ectomycorrhizae.

Studies have repeatedly shown that plant growth is enhanced by the presence of mycorrhizae. Mycorrhizae are particularly abundant in forest soils but are found in almost all soils, with the possible exception of grasslands where no trees have previously grown. Growth enhancement is especially significant for plants growing on infertile soils and dry soils. Interestingly, mycorrhizae development decreases following heavy fertilisation of soil.

Bacteria

The most abundant organisms in the soil are bacteria. Bacteria are minuscule, one-celled organisms that can only be seen with a powerful light (100X) or electron microscope. They can be so numerous that a pinch of soil can contain millions of organisms. Soils often have between 1,000,000 to 10,000,000 bacteria per gram. Bacteria are tough, they occur everywhere on earth and have even been found over a mile down into the core of the earth. Bacteria have an extremely varied metabolism. Bacteria can use reduced inorganics, the sun, or organics as an energy source. Some bacteria can live without free molecular oxygen.

Bacteria are common throughout the soil, but tend to be most abundant in or adjacent to plant roots, an important food source. Actinomycetes are a broad group of bacteria that form thread-like filaments in the soil. They are responsible for the distinctive scent of freshly exposed, moist soil. Actinomycetes are particularly effective at breaking down tough substances like cellulose (which makes up the cell walls of plants) and chitin (which makes up the cell walls of fungi) even under harsh conditions, such as high soil pH.

Free-living bacteria fix atmospheric nitrogen, adding it to the soil nitrogen pool. Other nitrogen-fixing bacteria form associations with the roots of leguminous plants such as lupine, clover, alfalfa, and milk vetch. Actinomycetes form associations with some non-leguminous plants (important species are bitterbrush, mountain mahogany, cliff rose, and ceanothus) and fix nitrogen, which is then available to both the host and other plants in the near vicinity. Some bacteria exude a sticky substance that helps bind soil particles into small aggregates. So despite their small size, they help improve water infiltration, water- holding capacity, soil stability, and aeration.

Bacteria are becoming increasingly important in bioremediation, meaning that we (people) can use bacteria to help us clean up our messes. Bacteria are capable of filtering and degrading a large variety of human-made pollutants in the soil and groundwater so that they are no longer toxic. The list of materials they can detoxify includes herbicides, heavy metals, and petroleum products. Bacteria can be divided into two large groups based on their carbon source. Autotrophic bacteria are independent of any carbon in the soil since they fix atmospheric CO_2; and obtain energy from the reactions of nitrogen and sulphur compounds in the soil.

Heterotrophic bacteria require carbon compounds as a food source. They are extremely important in decomposing organic matter. In the carbon cycle, CO_2; is absorbed by plants during photosynthesis. As these plants die and are incorporated into the soil, heterotrophic bacteria and other soil organisms decompose this organic matter and release CO_2; into the soil atmosphere. This movement of CO_2 into the atmosphere completes the carbon cycle.

The living organisms in the soil represent the key to plant health and to human health. Soil microorganisms are the essential link between mineral reserves and plant growth. Ecological soil management aims at assisting all soil organisms not substituting for them with a chemical system.

Rhizobia

Rhizobia are bacteria which can grow and multiply on artificial food sources. They are very difficult to identify from their shape or size alone because all forms are extremely small, short rods with rounded ends. Recently, new methods of rhizobia characterisation have been developed using DNA patterns from known rhizobia grown under laboratory conditions. Comparisons are made between sections of the known DNA molecule that code for particular functions (*e.g.* nitrogen fixation) with samples of DNA from soil or from a root nodule. This tells the scientists whether there are genes for that function in the soil sample or root nodule.

Serological techniques can also be used to identify rhizobia. Samples of known bacteria are injected into an animal under specific laboratory conditions. The animal's body produces antibodies against the bacteria. (A process that is similar to vaccination). The component of the blood containing the antibodies is the serum and this is separated from the red blood cells. This is called antiserum. The antiserum produced by the animal can be collected and used to help locate bacteria in a sample.

The antibodies in the antiserum can be joined to a fluorescent dye or to an enzyme with an attached dye. When the antibody is added to the corresponding bacteria it attaches firmly. The bacteria can then be identified using a fluorescence microscope or in a solution that contains a molecule that can interact with the enzyme and cause a colour change in the solution. This method is called the ELISA method. The antiserum can be chosen to be very specific for particular types of rhizobia, so it is a good way to identify whether those rhizobia are present in a root nodule. It is often important to know whether the rhizobia that were added with the seed have actually been responsible for forming the nodules on the legume or whether other rhizobia already in the soil formed the nodules.

Actinomycetes

Actinomycetes are thread-like bacteria that look like fungi. While not as numerous as bacteria, they too perform vital roles in the soil. Like the bacteria, they help decompose organic matter into humus, releasing nutrients. They also produce antibiotics to fight diseases of roots. Many of these same antibiotics are used to treat human diseases. Actinomycetes are responsible for the sweet, earthy smell noticed whenever a biologically active soil is tilled.

Algae

Many different species of algae live in the upper half-inch of the soil. Unlike most other soil organisms, algae produce their own food through photosynthesis. They appear as a greenish film on the soil surface following a saturating rain. Algae improve soil structure by producing slimy substances that glue soil together into water-stable aggregates. Some species of algae (the blue-greens) can fix their own nitrogen, some of which is later released to plant roots.

Protozoa

Protozoa are free-living microorganisms that crawl or swim in the water between soil particles. Many soil protozoa are predatory, eating other microbes. One of the most common is an amoeba that eats bacteria. By eating and digesting bacteria, protozoa speed up the cycling of nitrogen from the bacteria, making it more available to plants.

Nematodes

Nematodes are abundant in most soils, and only a few species are harmful to plants. The harmless species eat decaying plant litter, bacteria, fungi, algae, protozoa, and other nematodes. Like other soil predators, nematodes speed the rate of nutrient cycling.

Soil Organisms and Quality

All these organisms—from the tiny bacteria up to the large earthworms and insects—interact with one another in a multitude of ways in the soil ecosystem. Organisms not directly involved in decomposing plant wastes may feed on each other or each other's waste products or the other substances they release. Among the substances released by the various microbes are vitamins, amino acids, sugars, antibiotics, gums, and waxes.

Research on life in the soil has determined that there are ideal ratios for certain key organisms in highly productive soils. Roots can also release into the soil various substances that stimulate soil microbes. These substances serve as food for select organisms. Some scientists and practitioners theorise that plants use this means to stimulate the specific population of microorganisms capable of releasing or otherwise producing the kind of nutrition needed by the plants. Research on life in the soil has determined that there are ideal ratios for certain key organisms in highly productive soils.

Understanding the role that soil organisms play is critical to sustainable soil management. Based on that understanding, focus can be directed toward strategies that build both the numbers and the diversity of soil organisms. Like cattle and other farm animals, soil livestock require proper feed. That feed comes in the form of organic matter. Like cattle and other farm animals, soil livestock require proper feed. Organic matter and humus are terms that describe somewhat different but related things.

Organic matter refers to the fraction of the soil that is composed of both living organisms and once-living residues in various stages of decomposition. Humus is only a small portion of the organic matter. It is the end product of organic matter decomposition and is relatively stable. Further decomposition of humus occurs very slowly in both agricultural and natural settings. In natural systems, a balance is reached between the amount of humus formation and the amount of humus decay. This balance also occurs in most agricultural soils, but often at a much lower level of soil humus.

Humus contributes to well-structured soil that, in turn, produces high-quality plants. It is clear that management of organic matter and humus is essential to sustaining the whole soil ecosystem. The benefits of a topsoil rich in organic matter and humus are many. They include rapid decomposition of crop residues, granulation of soil into water-stable aggregates, decreased crusting and clodding, improved internal drainage, better water infiltration, and increased water and nutrient holding capacity.

Improvements in the soil's physical structure facilitate easier tillage, increased water storage capacity, reduced erosion, better formation and harvesting of root crops, and deeper, more prolific plant root systems. Soil organic matter can be compared to a bank account for plant nutrients. Soil containing 4% organic matter in the top seven inches has 80,000 pounds of organic matter per acre. That 80,000 pounds of organic matter will contain about 5.25% nitrogen, amounting to 4,200 pounds of nitrogen per acre. Assuming a 5% release rate during the growing season, the organic matter could supply 210 pounds of nitrogen to a crop. However, if the organic matter is allowed to degrade and lose nitrogen, purchased fertiliser will be necessary to prop up crop yields.

All the soil organisms mentioned previously, except algae, depend on organic matter as their food source. Therefore, to maintain their populations, organic matter must be renewed from plants growing on the soil, or from animal manure, compost, or other materials imported from off site. When soil livestock are fed, fertility is built up in the soil, and the soil will feed the plants. Ultimately, building organic matter and humus levels in the soil is a matter of managing the soil's living organisms—something akin to wildlife management or animal husbandry. This entails working to maintain favorable conditions of moisture, temperature, nutrients, pH, and aeration. It also involves providing a steady food source of raw organic material.

Soil Tilth

A soil that drains well, does not crust, takes in water rapidly, and does not make clods is said to have good tilth. Tilth is the physical condition of the soil as it relates to tillage ease, seedbed quality, easy seedling emergence, and deep root penetration. Good tilth is dependent on aggregation—the process whereby individual soil particles are joined into clusters or "aggregates." Ultimately, building organic matter and humus in the soil is a matter of managing the soil's living organisms.

Aggregates form in soils when individual soil particles are oriented and brought together through the physical forces of wetting and drying or freezing and thawing. Weak electrical forces from calcium and magnesium hold soil particles together when the soil dries. When these aggregates become wet again, however, their stability is challenged, and they may break apart. Aggregates can also be held together by plant roots, earthworm activity, and by glue-like products produced by soil microorganisms. Earthworm-created aggregates are stable once they come out of the worm. An aggregate formed by physical forces can be bound together by fine root hairs or threads produced by fungi.

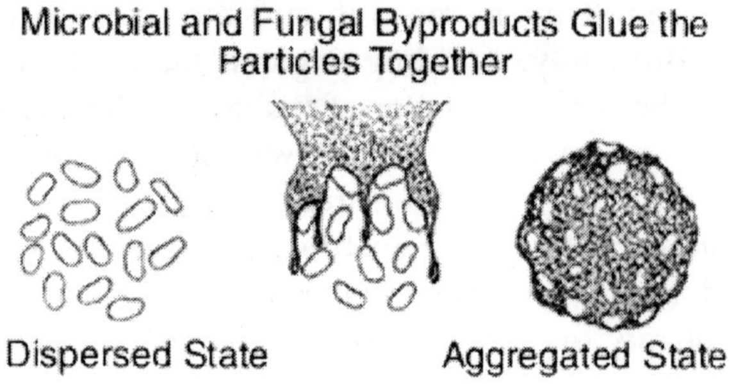

Figure 1. Microbial byproducts glue soil particles into water-stable aggregates.

Aggregates can also become stabilised (remain intact when wet) through the by-products of organic matter decomposition by fungi and bacteria—chiefly gums, waxes, and other glue-like substances. These by-products cement the soil particles together, forming water-stable aggregates (Figure 1). The aggregate is then strong enough to hold together when wet—hence the term "water-stable."

A well-aggregated soil allows for increased water entry, increased air flow, and increased water-holding capacity. Plant roots occupy a larger volume of well-aggregated soil, high in organic matter, as compared to a finely pulverised and dispersed soil, low in organic matter. Roots, earthworms, and soil arthropods can pass more easily through a well-aggregated soil. Aggregated soils also prevent crusting of the soil surface. Finally, well-aggregated soils are more erosion resistant, because aggregates are much heavier than their particle components.

The opposite of aggregation is dispersion. In a dispersed soil, each individual soil particle is free to blow away with the wind or wash away with overland flow of water. Clay soils with poor aggregation tend to be sticky when wet, and cloddy when dry. If the clay particles in these soils can be aggregated together, better aeration and water infiltration will result. Sandy soils can benefit from aggregation by having a small amount of dispersed clay that tends to stick between the sand particles and slow the downward movement of water.

Crusting is a common problem on soils that are poorly aggregated. Crusting results chiefly from the impact of falling raindrops. Rainfall causes clay particles on the soil surface to disperse and clog the pores immediately beneath the surface. Following drying, a sealed soil surface results in which most of the pore space has been drastically reduced due to clogging from dispersed clay particles. Subsequent rainfall is much more likely to run off than to flow into the soil (Figure 2). Since raindrops start crusting, any management practices that protect the soil from their impact will decrease crusting and increase water flow into the soil.

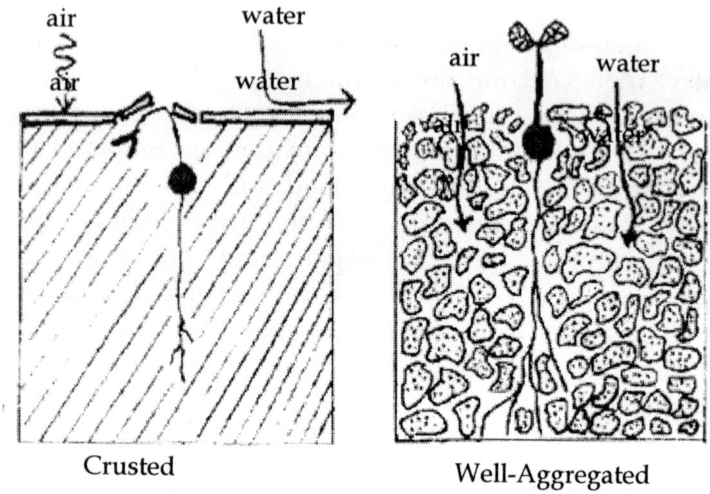

Figure 2. Effects of aggregation on water and air entry into the soil.

Mulches and cover crops serve this purpose well, as do no-till practices, which allow the accumulation of surface residue. Also, a well-aggregated soil will resist crusting because the water-stable aggregates are less likely to break apart when a raindrop hits them. Long-term grass production produces the best-aggregated soils. A grass sod extends a mass of fine roots throughout the topsoil, contributing to the physical processes that help form aggregates.

Roots continually remove water from soil microsites, providing local wetting and drying effects that promote aggregation. Fine root hairs also bind soil aggregates together. Roots also produce food for soil microorganisms and earthworms, which in turn generate compounds that bind soil particles into water-stable aggregates. In addition, perennial grass sods provide protection from raindrops and erosion. Thus, a perennial cover creates a combination of conditions optimal for the creation and maintenance of well-aggregated soil.

Conversely, cropping sequences that involve annual plants and extensive cultivation provide less vegetative cover and organic matter, and usually result in a rapid decline in soil aggregation. The best-aggregated soils are those that have been in long-term grass production. Farming practices can be geared to conserve and promote soil aggregation. Because the binding substances are themselves susceptible to microbial degradation, organic matter needs to be replenished to maintain microbial populations and overall aggregated soil status. Practices should conserve aggregates once they are formed, by minimising factors that degrade and destroy aggregation. Some factors that destroy or degrade soil aggregates are:

- bare soil surface exposed to the impact of raindrops,
- removal of organic matter through crop production and harvest without return of organic matter to the soil,
- excessive tillage,
- working the soil when it is too wet or too dry,
- use of anhydrous ammonia, which speeds up decomposition of organic matter,
- excess nitrogen fertilisation,
- allowing the build-up of excess sodium from irrigation or sodium-containing fertilisers.

Several factors affect the level of organic matter that can be maintained in a soil. Among these are organic matter additions, moisture, temperature, tillage, nitrogen levels, cropping, and fertilisation. The level of organic matter present in the soil is a direct function of how much organic material is being produced or added to the soil versus the rate of decomposition. Achieving this balance entails slowing the speed of organic matter decomposition, while increasing the supply of organic materials produced on site and/or added from off site. Moisture and temperature also profoundly affect soil organic matter levels.

High rainfall and temperature promote rapid plant growth, but these conditions are also favorable to rapid organic matter decomposition and loss. Low rainfall or low temperatures slow both plant growth and organic matter decomposition. The native Midwest prairie soils originally had a high amount of organic matter from the continuous growth and decomposition of perennial grasses, combined with a moderate temperature that did not allow for rapid decomposition of organic matter. Moist and hot tropical areas may appear lush because of rapid plant growth, but soils in these areas are low in nutrients. Rapid decomposition of organic matter returns nutrients back to the soil, where they are almost immediately taken up by rapidly growing plants.

Tillage can be beneficial or harmful to a biologically active soil, depending on what type of tillage is used and when it is done. Tillage affects both erosion rates and soil organic matter decomposition rates. Tillage can reduce the organic matter level in croplands below 1%, rendering them biologically dead. Clean tillage involving moldboard plowing and disking breaks down soil aggregates and leaves the soil prone to erosion from wind and water.

The moldboard plow can bury crop residue and topsoil to a depth of 14 inches. At this depth, the oxygen level in the soil is so low that decomposition cannot proceed adequately. Surface-dwelling decomposer organisms suddenly find themselves suffocated and soon die. Crop residues that were originally on the surface but now have been turned under will putrefy in the oxygen-deprived zone. This rotting activity may give a putrid smell to the soil. Furthermore, the top few inches of the field are now often covered with subsoil having very little organic matter content and, therefore, limited ability to support productive crop growth.

The topsoil is where the biological activity happens—it's where the oxygen is. That's why a fence post rots off at the surface. In terms of organic matter, tillage is similar to opening the air vents on a wood-burning stove; adding organic matter is like adding wood to the stove. Ideally, organic matter decomposition should proceed as an efficient burn of the "wood" to release nutrients and carbohydrates to the soil organisms and create stable humus. Shallow tillage incorporates residue and speeds the decomposition of organic matter by adding oxygen that microbes need to become more active.

In cold climates with a long dormant season, light tillage of a heavy residue may be beneficial; in warmer climates it is hard enough to maintain organic matter levels without any tillage. Tillage also reduces the rate of water entry into the soil by removal of ground cover and destruction of aggregates, resulting in compaction and crusting. Table 2 shows three different tillage methods and how they affect water entry into the soil. Notice the direct relationship between tillage type, ground cover, and water infiltration. No-till has more than three times the water infiltration of the moldboard-plowed soil.

Additionally, no-till fields will have higher aggregation from the organic matter decomposition on site. The surface mulch typical of no-till fields acts as a protective skin for the soil. This soil skin reduces the impact of raindrops and buffers the soil from temperature extremes as well as reducing water evaporation.

Table 2. Tillage effects on water infiltration and ground cover.

	Water Infiltrationmm/ minute	Ground CoverPercent
No-till	2.7	48
Chisel Plow	1.3	27
Moldboard Plow	0.8	12

Both no-till and reduced-tillage systems provide benefits to the soil. The advantages of a no-till system include superior soil conservation, moisture conservation, reduced water runoff, long-term buildup of organic matter, and increased water infiltration. A soil managed without tillage relies on soil organisms to take over the job of plant residue incorporation formerly done by tillage. On the down side, no-till can foster a reliance on herbicides to control weeds and can lead to soil compaction from the traffic of heavy equipment.

Other conservation tillage systems include ridge tillage, minimum tillage, zone tillage, and reduced tillage, each possessing some of the advantages of both conventional till and no-till. These systems represent intermediate tillage systems, allowing more flexibility than either a no-till or conventional till system might. They are more beneficial to soil organisms than a conventional clean-tillage system of moldboard plowing and disking. Adding manure and compost is a recognised means for improving soil organic matter and humus levels. In their absence, perennial grass is the only crop that can regenerate and increase soil humus.

Cool-season grasses build soil organic matter faster than warm-season grasses because they are growing much longer during a given year. When the soil is warm enough for soil organisms to decompose organic matter, cool-season grass is growing. While growing, it is producing organic matter and cycling minerals from the decomposing organic matter in the soil. In other words, there is a net gain of organic matter because the cool-season grass is producing organic matter faster than it is being used up.

With warm-season grasses, organic matter production during the growing season can be slowed during the long dormant season from fall through early spring. During the beginning and end of this dormant period, the soil is still biologically active, yet no grass growth is proceeding. Some net accumulation of organic matter can occur under warm-season grasses, however. In a Texas study, switchgrass (a warm-season grass) grown for four years increased soil carbon content from 1.1% to 1.5% in the top 12 inches of soil. In hot and moist regions, a cropping rotation that includes several years of pasture will be most beneficial.

Impacts of Nitrogen

Excessive nitrogen applications stimulate increased microbial activity, which in turn speeds organic matter decomposition. The extra nitrogen narrows the ratio of carbon to nitrogen in the soil. Native or uncultivated soils have approximately 12 parts of carbon to each part of nitrogen, or a C:N ratio of 12:1. At this ratio, populations of decay bacteria are kept at a stable level, since additional growth in their population is limited by a lack of nitrogen. When large amounts of inorganic nitrogen are added, the C:N ratio is reduced, which allows the populations of decay organisms to explode as they decompose more organic matter with the now abundant nitrogen.

While soil bacteria can efficiently use moderate applications of inorganic nitrogen accompanied by organic amendments (carbon), excess nitrogen results in decomposition of existing organic matter at a rapid rate. Eventually, soil carbon content may be reduced to a level where the bacterial populations are on a starvation diet. With little carbon available, bacterial populations shrink, and less of the free soil nitrogen is absorbed. Thereafter, applied nitrogen, rather than being cycled through microbial organisms and re-released to plants slowly over time, becomes subject to leaching. This can greatly reduce the efficiency of fertilisation and lead to environmental problems.

Excessive nitrogen stimulates increased microbial activity, which in turn speeds organic matter decomposition. To minimise the fast decomposition of soil organic matter, carbon should be added with nitrogen. Typical carbon sources—such as green manures, animal manure, and compost—serve this purpose well. Amendments containing too high a carbon to nitrogen ratio (25:1 or more) can tip the balance the other way, resulting in nitrogen being tied up in an unavailable form.

Soil organisms consume all the nitrogen in an effort to decompose the abundant carbon; tied up in the soil organisms, nitrogen remains unavailable for plant uptake. As soon as a soil microorganism dies and decomposes, its nitrogen is consumed by another soil organism, until the balance between carbon and nitrogen is achieved again.

Fertilisers

Commercial fertiliser can be a valuable resource to

farmers in transition to a more sustainable system and can help meet nutrient needs during times of high crop nutrient demand or when weather conditions result in slow nutrient release from organic resources. Commercial fertilisers have the advantage of supplying plants with immediately available forms of nutrients. They are often less expensive and less bulky to apply than many natural fertilisers.

Not all conventional fertilisers are alike. Many appear harmless to soil livestock, but some are not. Anhydrous ammonia contains approximately 82% nitrogen and is applied subsurface as a gas. Anhydrous speeds the decomposition of organic matter in the soil, leaving the soil more compact as a result. The addition of anhydrous causes increased acidity in the soil, requiring 148 pounds of lime to neutralise 100 pounds of anhydrous ammonia, or 1.8 pounds of lime for every pound of nitrogen contained in the anhydrous.

Anhydrous ammonia initially kills many soil microorganisms in the application zone. Bacteria and actinomycetes recover within one to two weeks to levels higher than those prior to treatment. Soil fungi, however, may take seven weeks to recover. During the recovery time, bacteria are stimulated to grow more, and decompose more organic matter, by the high soil nitrogen content. As a result, their numbers increase after anhydrous applications, then decline as available soil organic matter is depleted. Farmers commonly report that the long-term use of synthetic fertilisers, especially anhydrous ammonia, leads to soil compaction and poor tilth. When bacterial populations and soil organic matter decrease, aggregation declines, because existing glues that stick soil particles together are degraded, and no other glues are being produced.

Potassium chloride (KCl) (0-0-60 and 0-0-50), also known as muriate of potash, contains approximately 50 to 60% potassium and 47.5% chloride. Muriate of potash is made by refining potassium chloride ore, which is a mixture of potassium and sodium salts and clay from the brines of dying lakes and seas. The potential harmful effects from KCl can be surmised from the salt concentration of the material. Some plants such as tobacco, potatoes, peaches, and some legumes are especially sensitive to chloride.

High rates of KCl must be avoided on such crops. Potassium sulphate, potassium nitrate, sul-po-mag, or organic sources of potassium may be considered as alternatives to KCl for fertilisation. Sodium nitrate, also known as Chilean nitrate or nitrate of soda, is another high-salt fertiliser. Because of the relatively low nitrogen content of sodium nitrate, a high amount of sodium is added to the soil when normal applications of nitrogen are made with this material. The concern is that excessive sodium acts as a dispersant of soil particles, degrading aggregation.

4

Soil Microbiology

The main sampling problems in soil microbiology are usually a result of the complexity of the medium being sampled. If the type of method being used requires a "generalised" sample of the soil, the problem is to determine what soil horizons to sample, how many samples to take, exactly where to take the samples (to determine spatial variation), how often to take the samples (to determine temporal variation) and what size of samples should be used.

Methods Used in Microbial Ecology

1. Detection of growth habit, form and pattern arrangement in soil,
2. Isolation of specific groups of microorganisms or entire microbiota,
3. Detection and measurement of activities in soil,
4. Measurement of biomass.

Detection of Growth Habit, Form and Pattern or Arrangement in Soil

Microorganisms can have many different growth patterns in soil. Some bacteria form microcolonies on the surfaces of mineral or organic particles, some fungi and Actinomycetes form visible and distinct diffuse colony forms over many soil particles, some bacteria exist as single cells and others form colonial structures. These different morphologies may be at many different scales—some fungal colony structures can be kilometres in extent, while some bacterial colonies an exist on a clay particle.

In addition, individual cells in one of these structures can vary in morphology depending on the stage in the growth cycle, nutrient status, physical conditions, etc. Arthobacter species often exhibit polymorphism—the cell shape changes based on its growth rate and age. They range from coccoid cells to bacilli. It is often necessary to associate microorganisms with other structures or objects in soil. Plant roots, organic materials, mineral grains, fungal mycelium and arthropods are all colonised by particular types of bacteria and Actinomycetes.

Many types of microbial activities are only possible under certain conditions. The distribution of the cells in soil often reflects their activities on certain substrates or their responses to chemical or physical conditions. Some anaerobic cells such as Clostridium will only be found in anaerobic regions. Formation of resistant spores is a common response to adverse conditions and can be used to indicate previous growth habits or arrangements.

Using standard techniques, it is usually impossible to both locate spatially and also identify a microorganism in a soil sample. To identify microorganisms, the cells are usually cultured in some way to permit colony formation for subsequent identification. Techniques to detect form, pattern and arrangement in soil can be broken down into:

1. Microscopic methods:
   - Light microscopy,

- Electron microscopy.

2. Microscopic methods plus culturing:
   - Direct culture methods,
   - Fluorescent antibody and related methods,
   - Miscellaneous methods.

Microscopic methods

Light microscopy: Many older methods using direct microscopic examination of soil samples are still in use today because of their simplicity. They are especially useful when examining smaller soil samples such as pieces of organic materials or mineral grains. There are two main types of methods used to visualise the microorganisms in these samples; classical stains such as phenol aniline blue and fluorescent stains such as fluorescein isothiocyanate.

The first can be examined after staining with any bright-field, white light microscope assuming that light can be transmitted through the object being examined.

The second uses a stain that emits light at a visible wavelength when illuminated with far-violet or ultraviolet light. This can be incident illumination that does not have to pass through the object.

The mercury arc lamp is a strong source of ultraviolet light that is filtered though an excitation filter and passed to a dichroic mirror. This mirror is coated with a very thin metal film that reflects ultraviolet light but does not allow it to pass through. It does allow visible light to go through and not be reflected. The ultraviolet light is passed down through the objective lens and focused onto the object from the top. If the fluorescent dyes staining the microorganisms fluoresce in the visible spectrum, the emitted light is collected by the objective lens and passes through the dichroic mirror to the eyepiece lens and the eye of the observer. The eyepieces always contain a barrier filter that prevents any of the ultraviolet light from reaching the eyes of the observer.

The most common fluorescent stains are acridine orange, fluorescein isothiocyanate (FITC) and rhodamine (fluoresces red). They both react with parts of protein molecules—the sulfhydryl groups—and attach strongly to the protein molecules. Another example is Calcofluor; it reacts with cellulose, chitin and similar polysaccharides and it useful for staining fungi and Actinomycetes. It is also relatively non-toxic and can be used as a vital stain to examine living cells. Other fluorescent stains include europium chelate (europium (iii) thenoyltrifluoroacetonate) that stains nucleic acids and DAPI (4′-6′-diamidino-2-phenyl-indole), ethidium bromide and Hoechst 33258 (bisbenzimide) that all stain DNA.

Another group of fluorescent stains are the fluorescent probes. They differ from FITC and rhodamine in that they are not fluorescent until they come into contact with the correct environment. Typically this is the lipid within microbial cells. Only then do they fluoresce and emit visible light. Examples of this group are DANSYL chloride and the 8-anilino-1-naphthalene sulfonic acid salts (Mg-ANS and Na-ANS). Their major advantage is that they can be applied to soil samples and immediately examined without removing excess, unreacted stain. FITC and rhodamine need extensive washing to remove unreacted stain.

Electron microscopy: Electron microscopy can be used in a manner similar to light microscopy to examine very small areas within soils. Both scanning and transmission electron microscopy has been used, but the major disadvantage of both is the effort required to examine even relatively small volumes of soil. Another problem is the fixation and coating/mounting processes involved. They often destroy some of the features of the microbial cells. The process is tedious and can only examine very small regions of a soil sample.

Microscopic methods plus culturing

Soil samples can be impregnated with agar or polyacrylate resins, sectioned into thin "plates" Soil and examined by direct microscopy. Diamond microtomes are needed to section through mineral grains but the resulting very thin sections can be examined by transmission light microscopy.

Direct culture methods: If soil can be removed intact, placed on a nutrient medium and incubated, the resulting small colonies can be examined macroscopically or microscopically. Using selective media, particular types of microorganisms can be selectively cultured and identified. One technique uses Scotch tape to take successive samples from an exposed soil surface. The samples are transferred to a selective agar medium and incubated. The position of colonies on the plate reveals the location and distribution of the original cells in the soil. This technique can be combined with replica plating of the agar plates to test the reactions of the colonies to different media thus aiding in identification of the colonies.

Fluorescent antibody and related methods (immunofluorescence methods): The fluorescent antibody technique is the only one that can simultaneously locate and identify microorganisms in intact soil samples or sections.

The antibodies to microbial cells are produced by injecting the cells under study into a suitable animal. After incubation, the animals produce antibodies to the microbial cells that can be isolated from serum samples of the animals. The antibodies are proteins so can be reacted with FITC to produce FITC-antibody conjugates. These FITC-antibodies will only adhere to the correct microbial cells if applied to a soil sample. When excess FITC-antibody conjugate has been removed by washing, only those microbial cells will fluoresce and they can be simultaneously located and identified by epifluorescence microscopy.

This technique has been extensively used in medical microbiology to identify Neisseria gonorrhoeae, enteropathogenic *Escherichia coli, Salmonella spp., beta-haemolytic streptococci* and viruses in tissue samples. It has been used in soil microbiology to identify nitrogen-fixing *Rhizobium* spp., *Bacillus* spp., various fungal genera such as Aspergillus, and a few Actinomycetes. One problem is the relatively non-specific nature of many antibody preparations. Many bacteria in the same general taxonomic group have similar chemical structures on their cells and produce a complex of antibodies from those structures that overlap with the complex produced by the other similar bacteria.

Thus if the antibody complex is used to form the conjugate with FITC, that FITC-Ab will cross-react with the other bacteria as well as the target species. Usually this reaction will be weaker but still significant. One way to "purify" the complex is to remove the cross-reacting antibodies by reacting them with the actual bacterial cells from the unwanted cross-reacting species.

Any common antibodies will be adsorbed onto the surfaces of the added cells and removed from the complex. The resulting remaining antibodies will be much more specific to the target species. A more recent modification of the method uses monoclonal or polyclonal antibodies produced in other microbial cells to obtain larger quantities of antibody for conjugation with FITC. Many of these antibodies are now available commercially from suppliers and some are available already conjugated and therefore labelled with FITC and/ or rhodamine.

Enzyme-linked immunosorbant assays: ELISA assays have found some use in soils when populations sought for exceed 10,000 cells per ml. The technique has been applied mainly to Rhizobia in soil and roots of legumes. The major difficulty is removal of the microbial cells from the substrates; both direct lysis in situ and removal of cells followed by lysis have been used. Gene probe and nucleic acid hybridisation techniques rely on detecting specific sequences of nucleic acids in the organisms under study.

If the sequences used are carefully chosen to be diagnostic, this technique can find specific organisms in soils and other environmental samples. The gene probe is a short segment of nucleotides that binds specifically with the homologous sequence in the target microorganism. If the segment is labelled with radioactive ^{32}P, any binding to the target nucleotides can be detected by the presence of the radioactivity after reaction. The Polymerase Chain Reaction (PCR) has recently been applied to soil microbial ecology.

In this technique, extracted DNA is melted to form single strands, annealed with primers, and the DNA is extended from the primers by nucleotide addition using DNA polymerase enzyme. The primers are chosen to link to regions of DNA of interest (close to a diagnostic target sequence). Bioluminescence marker genes, typically the lux gene of Vibrio fischeri. This gene causes photoluminescence in the bacterium.

If the gene can be inserted into the target organisms, they become photoluminescent and this property can be used to detect them and track their fate in soil and water samples. This technique has been used with E. coli, and Pseuodomonas target organisms. It has been extended by fusing other genes with the lux gene and inserting both into cells. Naphthalene degradation is promoted by a gene (nah) which has been combined with the !ux gene to make a diagnostic pair. This can track both the bacteria and their activity in soil samples.

Miscellaneous methods: Various methods have been used by individual investigators to determine microbial distribution in soils. Rossi-Cholodny slides are simple microscope slides buried in soil and left for various times before microscopic examination. Strips of transparent chitin and cellulose materials have also been used. Pedoscopes are small-bore glass capillary tubes used in a similar manner. A few workers have used direct micromanipulation of soil under a microscope to remove and incubate individual pieces of soil or even individual microorganisms.

Using that technique, it was shown that the basidiomycete fungi were much more common than previously thought, They are under-represented in dilution plates because of the rapid growth of spores from other fungi. When pieces of fungal mycelium were isolated directly using micromanipulation tools, the basidiomycetes were much more numerous.

Isolation of Microorganisms

– Elective culture methods,

– Selective culture and media, and
– Non-selective media and methods.

Elective culture methods

If specific chemical compounds are added to soil in the field or in the laboratory and incubated under specific conditions, the organisms capable of growing under those conditions will multiply and will come to comprise a greater percentage of the total microbiota. Alternatively, if specific inhibitory chemicals or incubation conditions are used specific parts of the microbiota will be decreased in overall percentage. This very simple concept is the basis for a large fraction of the industrial uses for soil microbiology such as isolating hydrocarbon-degrading bacteria, isolating bacteria that degrade various pollutants such as pesticides, PCBs, organochlorines, isolating bacteria or fungi that produce new antibiotics, and many other examples. Because of the diversity of the microbiota in soil, this technique can often be applied to isolate microorganisms with any desired property.

For example, to isolate a bacterial isolate that can degrade cellulose under acid, anaerobic conditions, it is only necessary to add cellulose (cotton, paper, etc.) to a soil sample that has been acidified (with acetic acid, addition of sulphur, etc.) and incubate it anaerobically. A more sophisticated approach might be to find a soil that has been naturally subjected to conditions similar to those that you wish to use and use that as a sample source. Thus a peat bog would have a microbial population already adapted to acidic and anaerobic conditions with the presence of cellulose.

Using that as a source of material would probably lead to more successful isolations. In a similar manner, soil from an oil refinery contains more hydrocarbon degrading bacteria than farm field soils. Despite this, it is often found that many soils contain a very large variety of organisms that do not seem to be related to that soils obvious chemical or physical environment; most farm field soils do contain hydrocarbon degrading bacteria. This could be a reflection of the large degree of diversity in soil leading to many microbial niches, a result of the biochemical diversity even within individual species of microorganisms, or movement of bacteria through wind or water transport mechanisms.

Selective culture and media

All media used in microbiological laboratories are selective to some degree or another; there are no truly non-selective media. It is possible to make media and incubation conditions very selective by chemical and physical modifications. This is often very useful to isolate and count particular groups of bacteria or other organisms from soil samples. There are routine ways to produce selective media:

(a) Add compounds used by an organism as a nutrient source.

(b) Omit compounds required by most other organisms (omit nitrates and other fixed nitrogen sources to isolate nitrogen fixing bacteria).

(c) Add selectively biocidal compounds (penicillin to inhibit gram-positive bacteria, neomycin and streptomycin as general bacterial inhibitors, actidone and nystatin as general fungal inhibitors).

(d) Change physical properties (pH, pE [redox potential], etc.).

(e) Alter incubation conditions (temperature, water content, osmotic pressure, light, etc.).

Using combinations of these techniques it is possible to design very selective media; Pseudomonas isolation media is very specific for its named bacterial group. In theory, almost any physiological group of organisms can be cultured selectively. A single stage isolation from a soil sample can be changed to a multistage isolation process by replica plating. Individual colonies are transferred in their original orientation on the plate by pressing a pad of sterile velvet cloth onto the surface of the plate, removing a small sample of each colony and pressing it onto the surface of a fresh plate. This can be a different growth medium so that only a part of the original population is able to grow on the new medium. In this way, progressive selective media can be used to isolate bacteria with combinations of properties. To enhance the detection of the particular group of microorganism under study, it is also possible to improve the diagnostic precision of the media by using some properties of the organisms.

Non-selective media

So-called "non-selective" media are only media and incubation conditions designed to isolate as large a part of the microbiota in soil as is possible. Truly non-selective media do not exist. The least selective media today isolate maybe 1% to 10% of the total soil bacteria and maybe 5% to 15% of the fungal population of soils. The media used for bacteria and fungi are different. Some general observations:

- Fungi grow better on media with a high C:N ratio.
- Bacteria grow better on low C:N ratio media.
- More bacteria are isolated using neutral pH media.

- More fungi are isolated on low pH media.
- Spores from common fungi such as Aspergillus, Mucor and Penicillium in soil grow rapidly on many fungal isolation media and prevent growth of the slower-growing fungi.
- More bacteria are isolated on complex media (peptone, proteins, etc.).
- Many bacteria and fungi produce antibiotic compounds on plates and prevent growth of other microorganisms.
- Bacteria from soil or water tend to grow better on weaker strength media than do medically important bacteria. High nutrient levels can inhibit some soil or water isolates.
- Some bacteria are very difficult to culture at all on media in the laboratory. They are very slow growing or very sensitive to components of normal media.
- It is probable that there are bacteria that have never been cultured on media.

Detection and Measurement of Activities in Soil

Respiration measurements

Respiratory activity as a general activity measurement is well established. The Biometer flask is a common piece of apparatus used to measure the activity of soil samples. Soil is incubated at the required temperature, moisture content, etc. in the main flask and air can be added through the CO_2 absorber in the tube on top of the main flask. During incubation the CO_2 produced by respiration is absorbed into the alkali in the side tube. This is then titrated to discover the amount of alkali neutralised by the evolved CO_2.

More air can be added and the alkali replaced to continue the process of analysis. To obtain reproducible results from assays of respiration rate the soils are usually sieved to remove plant materials, insects, etc. This destroys aggregate structure and may stimulate activity due to release of nutrients previously protected inside the aggregates. Disturbance of any kind usually causes a temporary burst in respiratory activity.

Cell division rate

Cell division rates can be estimated by various counting methods during a given time. If in situ rates are required, the very slow growth rates of soil microorganisms makes this a very difficult measurement to obtain. Autoradiographic tracer experiments can be used to trace cell division rates. Nalidixic acid inhibits cell separation after cell growth so that applying this antibiotic to samples leads to long filamentous cells of bacteria. These indicate cell division rates. Vital stains such as ANS or Calcofluor can be used to trace microbial cell division.

Mycelial extension

Fungi grow by extension of mycelium; if this can be observed, it can be used to estimate growth rates. Direct observation or extraction from soil can be used.

Enzyme activity or content

The dehydrogenase enzymes can be extracted from soil samples, reacted with tetrazolium dyes to form coloured products and the colour measured by colorimetric methods. This gives an estimate of the dehydrogenase activity in the soil sample and thus an estimates of the organisms present. Various sieving techniques can be used to separate small soil animals, fungi and bacteria from soil samples, but they are all somewhat unsatisfactory—*i.e.* they do not completely separate the organisms even in a well-mixed soil suspension. Other enzymes that have been used to estimate activity include oxido-reductases, transferases, phospharases, cellulases, and proteases.

Substrate utilisation rate

Any compound that is a substrate for microorganisms can be used to estimate their activity if the decrease in substrate concentration can be measured. Examples are:

- carbohydrates—detected and measured by the anthrone reaction,
- amino acids—detected and measured by the ninhydrin reaction,
- proteins—detected and measured by the Biuret reaction (Folin) etc.

There are always problems of interference with assays, extraction problems, change in conditions as the substrates are used by the microorganisms, etc.

Product accumulation rate

Any product of microbial activity that can be detected can be used. Transitory products are difficult to assay. There are problems with both subtrate utilisation and product accumulation rate measurements:

1. As substrates disappear or products accumulate, conditions change (pH, Eh, CO_2 levels, O_2 levels, etc.)

2. Metabolites may be used in succession (sequentially) and removed at different rates depending upon conditions.

Radioisotope cycling studies

Many elements can be obtained in the form of a radioactive isotope (^{14}C, ^{32}P, ^{3}H [tritium] etc.). They can be used to measure the activities of microorganisms growing and metabolising compounds that include these isotopes. For example, glucose can be labelled with ^{14}C in specific positions in the glucose molecule and the fate of the radiolabelled atom can be traced through standard chemical analysis followed by detection of the radioactivity in specific fractions. If the radiolabelled atom has been evolved as CO_2 through respiratory of the cells, it will make CO_2 radioactive and this can be detected.

Autoradiography

If uniformly labelled ^{14}C-glucose is added to a soil and incubated, the rate of uptake of $^{14}C_2$ into cells can be detected using the technique of autoradiography. In this technique, the cells are added in suspension to the surface of a photographic emulsion on a plate. During incubation in the dark, the radioactivity in the cells causes the silver salts in the emulsion to become "exposed" and the silver grains produced show as dark spots on the emulsion. Presence of dark spots signals the presence of ^{14}C.

Any radioisotope can be used in this process to detect uptake. Different compounds can be purchased already labelled with specific radioisotopes of many different kinds. If the cell suspension or soil sample with added glucose is incubated longer, the glucose will be metabolised and can be detected after absorption in apparatus similar to that described under Respiration Measurements above. An organic liquid is used to absorb the CO_2 so that an organic scintillation liquid can be added directly to the absorbing fluid. This can then be placed in a scintillation counter and the flashes of light emitted by the scintillation fluid molecules as they are impacted by emitted particles from the decay of the radioisotope can be measured. In this manner, the amount of radioactivity can be measured.

Stable isotopes (*i.e.* non-radioactive) can be used in a similar manner to trace microbial activity). The incorporation of ^{3}H-labelled thymidine into DNA by cells is a popular measure of growth rates and activity.

Biomass Determination

A measure of the total microbial biomass in soils is often required when studying productivity or fertility of soils. Sometime the biomass of specific parts of the microbiota is require—for example fungal biomass versus bacterial biomass. If possible, the method should yield results comparable to the results used for plant and animal determinations so that the overall cycling of material in the system can be compared.

Early methods for biomass determination relied on counting microbial propagules and other cells directly by microscopy or by viable plate counting and converting these to biomass via some conversion factor. More recently, newer methods have been developed that use measures of components of all cells (*e.g.* ATP) to estimate biomass. These methods include:

1. *Soil fumigation method*: A sample of soil is sieved and placed in a container and CO_2 output measured over twenty days. There is typically a period of very rapid respiration followed by a much lower but stable respiration rate. If the organisms are killed by fumigation with chloroform, this initial flush of activity does not occur and the cells are killed. If the chloroform is removed there will be bigger flush or rapid respiration because dead microbial cells will contribute to the total substrates available. The difference between the normal rapid respiration and the greater amount after chloroform treatment is due to the amount of microbial biomass originally in the system.

2. *Determination of ATP content of soils*: ATP is extracted from the cells in the soil and measured by its reaction with the enzyme system luciferin + luciferinase. The enzyme luciferinase is extracted from firefly tails and emits light when ATP reacts with its substrate luciferin. The emitted light is measured in a scintillation counter. Specific machinery is available for this ATP determination that integrates the reaction/enzyme/substrate system into one vial of dedicated small scintillation counter. The amount of ATP per gram a cell material varies, but averages 10.0 moles g^{-1} resting biomass. It has been proposed that the ratio of ATP to biomass C content is 1:120 in soil samples and this is very close to the ratio in exponentially growing microorganisms and eucaryotic cells.

3. *Determination of cell wall components of bacteria*: Bacteria contain specific cell wall components such as muramic acid that can be released by acid hydolysis and analysed by High Pressure Liquid

Chromatography (HPLC). The amounts of muramic acid in bacterial cell walls varies depending on whether the cells are gram negative or gram positive. Thus, unless the proportion of Gram positive to Gram negative cells is known, this technique has serious limitations. Bacterial spores also contain up to four times the normal levels of muramic acid.

4. *Dilution plate counts and direct microscopical counting*: Dilution plates usually only are able to culture between 1 to 10% of the viable organisms in soil samples. Direct microscopic observation methods usually overestimate the number of cells because they include dead organisms or other particles in their count.

5

Microbial Communities in Soil

Sugarcane is vegetatively propagated. Therefore, sugarcane cultivars are clones composed of genetically identical plants. Buds on planted stalk sections germinate and grow to produce the first "plant cane" crop. Following harvest of the stalks, buds on the stubble remaining in the soil can germinate to produce a ratoon crop. Monoculture can lead to poor root health and growth constraints in many crops, including sugarcane.

Some researchers done a study on sugarcane fields in Louisiana to understand the microbial communities in this soils. Methods and the results of their study are described in this chapter.

Soilborne pathogens, such as *Pythium* and nematode species, have been shown to adversely affect sugarcane growth. However, evidence has suggested that additional biological factors may affect root health and contribute to yield decline in sugarcane. This possibility was suggested by results obtained by the following experimental methods:

1) biocide studies, in which selective fungicides, such as metalaxyl or nematicides, did not completely alleviate poor root growth, whereas fumigants and

dithiocarbamate broad-spectrum fungicides, mancozeb, maneb, and zineb, did, and

2) soil heat-treatments, in which soil solarisation increased plant growth.

Soil fumigation studies using methyl bromide, chloropicrin-methyl bromide or metham sodium have resulted in increased sugarcane growth and yield in fumigated soil compared to non-fumigated soil. Thus, fumigation has been used to benchmark potential yields and to reveal root health problems. The response to soil fumigation by many other crops, such as corn, barley, apples, tomatoes, and citrus indicate that soil-based growth constraint can occur in many crops. In addition, growth increases on previously non-cropped lands by crops, such as sugarcane, corn, citrus, peaches, and apples, indicate that the occurrence of soil-based growth constraints are widespread. In the past, culture methods involving the use of culture media designed to enhance the growth of many or selected species were used to investigate soil and rhizosphere microorganism communities.

It was later estimated that 0.01% to 10% of microorganisms in soil were capable of growth by cultivation. Therefore, a bias existed in soil microbial community studies because only microorganisms capable of growth under artificial conditions were being evaluated and not the total community. Count data obtained with fluorescence microscopy revealed that the majority of the non-colony-forming species in soil belonged to a group with very small cell size and high numbers and that the majority of the colonyforming species belonged to a group with large cell size and low numbers. It was suggested that the size of soil bacterial cells plays a role in their ability to grow in culture and that the ecological significance of colonyforming cells may be great despite their low numbers.

The researchers further suggested that the small size and slow metabolism of the noncolonyforming species implied that their energy requirement represented a small fraction of the energy in the soil ecosystem. As long as researchers remain aware of the limitations, selective media methods can be used to evaluate culturable species as a component of soil microbial diversity. To study the soil microbial communities associated with sugarcane rhizosphere and nonrhizosphere soils with a long-term sugarcane cropping history (SCH) and no recent sugarcane cropping history (NSCH), it would be desirable to compare quantitative and qualitative aspects of microorganisms isolated on a nonselective medium and different media selective for groups of organisms.

Soil dilution plating can establish countable colony numbers, and colonyforming units (cfu) per gram of soil can be calculated. A portion of the isolates can then be collected for identification to make qualitative comparisons of communities. For total culturable bacterial enumeration, 10% tryptic soy agar is widely used. Studies have shown that Pseudomonas species found in association with the rhizosphere can play either a beneficial or deleterious role in plant growth and crop yields. Pseudomonas species have played beneficial roles as biocontrol agents of rhizosphere associated phytopathogens, including Fusarium wilts on tomato, Rhizoctonia damping-off on cucumber roots, and *Gaeumannomyces graminis* take-all of wheat.

Some *Pseudomonas* spp. have been found to be plant pathogenic to lucerne, oats, and wheat. Nehl, et al. determined that Pseudomonas spp. along with other rhizobacteria can fluctuate from being deleterious to plant growth promoting depending on environmental conditions, host genotype and mycorrhizal status. Therefore, total and fluorescent culturable bacterial populations in Louisiana sugarcane soils were evaluated with two selective media: King's B and Pseudomonas agar were used. It has been reported that phytopathogenic *Pseudomonas* syringae strains produce siderophores, which complex with ferric iron to provide themselves with iron under iron limited conditions and to gain a competitive advantage over other microorganisms lacking these siderophores.

Kloepper, et. al. suggested that competition for iron is involved in biological control of plant diseases. In this case, beneficial *Pseudomonas* spp. produce extracellular siderophores which place the phytopathogens at a disadvantage because they are unable to acquire iron. Therefore, chrome-azurol sulphate medium was used in this study to enumerate culturable total and siderophore producing bacteria. Actinomycetes produce antibiotics that are antagonistic to a number of different microorganisms, including some plant pathogens; however, some actinomycetes are phytopathogens themselves. Therefore, total colonies developing on actinomycetes agar were enumerated and compared for soil with and without a sugarcane cropping history.

Other phytopathogenic fungi besides *Pythium* have been demonstrated to be associated with sugarcane roots, including *Pachymetra, Fusarium, Trichoderma, Rhizoctonia, Marasmius, Monilia, Collectotrichum, Phialophora, Humicola,* and *Thielaviopsis*. Conversely, some fungal species can play beneficial roles against root phytopathogens. *Trichoderma* has been found to be antagonistic to *Pythium* spp. in sugarcane soils, and Gliocladium virens has been shown to be antagonistic to Rhizoctonia solani in white beans. Therefore, rose-bengal streptomycin medium selective for fungi was used for fungal enumeration.

Fusarium species are common in agricultural soils. Many crops are damaged by Fusarium wilts. It has been demonstrated that Fusarium spp. are pathogenic on sugarcane. Therefore, Komada's medium selective for *Fusarium* was used to enumerate and compare fusarial fungi in this study. Phytopathogenic microorganisms can occur in root interiors. Therefore, to study the components of microbial communities colonising sugarcane roots in soils with a long-term sugarcane cropping history and no recent sugarcane cropping history, it would be desirable to compare the quantitative and qualitative aspects of microorganisms isolated from root interiors.

The Biolog™ bacterial identification system has been adapted to compare sole carbon substrate utilisation profiles (SCSUP) of microbial communities. Many studies have used SCSUP to examine soil microbial communities associated with roots of plants, including hybrid larch (*Larix eurolepis* A. Henry), Sitka spruce (*Picea sitchensis Bong. Carr.*), wheat (*Triticum* spp.), and grass species. In this study SCSUP was used to determine if sugarcane soils with and without a sugarcane cropping history could be distinguished from each other and to identify any carbon substrates responsible for the differences.

Biolog GN and GP Micro-Plates™ are 96 well microtitre plates that contain 95 different carbon substrates and one control well containing no carbon substrate. In the Biolog microplate, utilisation of a carbon source is indicated as a colourless tetrazolium redox dye is reduced to a coloured formazan product. Absorbance values of the colour response in each microplate well are recorded with an automated microplate reader. For community comparisons, the SCSUP data are collected from the microplate and can be analysed by multivariate techniques, including detrended correspondence analysis (DCA), canonical variate analysis (CVA), kinetic modelling or principal component analysis (PCA). These multivariate analyses discriminate communities based on their substrate utilisation patterns and reduce the dimensionality of the data set generated from the 95 substrates (variables) measured by absorbance values, to help determine the factors (variables) that contribute to the differences between the communities studied.

DCA uses weighted averages to transform data and is less sensitive to variation in average well cell densities in SCSUP. CVA requires a priori grouping of the variables in a study. Kinetic modelling requires frequent plate readings over several days and extra steps in manipulating data. PCA is the analysis of choice for most researchers using SCSUP because a priori grouping of the variables is not required, and it is less labour intensive. The SCSUP data also can be used to calculate different measures of microbial community functional diversity, such as substrate richness, substrate evenness, substrate diversity, and diversity and evenness. SCSUP has limitations.

The profiles only represent the activities of members of the community that grow rapidly under the conditions in the Biolog plates, and the ratio of numbers of replicates to the number of variables often used is not large enough to yield statistically stable results. However, Legendre and Legendre stated that the first few eigenvectors, those typically used in PCA, are little affected when the replicate-to-variable ratio is low; therefore, the problem of low replication with large numbers of variables should not lead to incorrect interpretation of ordination. If one is concerned about the low replicate-to-variable ratio, a method that decreases the number of variables and increases the number of replications per sample is now commercially available from Biolog.

The EcoPlate™ is a microtiter plate that is composed of 31 carbon substrates that are useful for soil community analysis, based on carbon sources previously requested by soil researchers. The 31 substrates are replicated three times on a single plate. However, carbon sources utilised in different soils vary; therefore, the extra substrate data obtained with the GN2 and GP2 MicroPlate™ system may be preferable to the EcoPlate™ in some applications. Another point that should be given consideration is whether to use the GN2, GP2, or both microplates. It has been reported that the rhizosphere environment is dominated by fastgrowing, nutritionally versatile gram-negative bacteria, thus the GN2 MicroPlate™ would produce more information than the GP2 MicroPlate™ in rhizosphere studies.

Also, 62 substrates are duplicated in the GN2 and GP2 plates with only 33 substrates unique to each plate. Therefore, the extra expense and work to gain information from just 33 extra GP2 substrates may not be justified. Most recently, molecular research methodologies capable of characterising microbial communities in soil have been developed. These methods depend on the efficient extraction of nucleic acids from the microorganisms of interest. There are numerous reports of different techniques to extract nucleic acids from soil microbial communities. The techniques lyse and extract nucleic acids either directly or indirectly from microorganisms.

With direct lysis methods, microbial cells are lysed within the soil matrix. With indirect methods, the microbial cells are separated from the soil matrix by centrifugation and then lysed to release nucleic acids. There are advantages and disadvantages involved with each technique. Direct lysis methods yield higher DNA concentrations and are presumably less biased than indirect lysis methods. However, a disadvantage to direct lysis methods is that more PCR-inhibitory substances are co-extracted along with the DNA. Numerous techniques are available to analyse extracted soil community nucleic acids.

Most produce profiles of the soil community based on PCR amplified DNA using universal primers. Some profiling methods are described below. Single-strand conformation polymorphism (SSCP) has been used to generate profiles of genes coding for 16S rRNA(DNA) based on minor differences in nucleotide sequence. Schmalenberger and Tebbe used SSCP to compare bacterial communities in rhizospheres of transgenic maize and sugar beet. PCR-based amplified ribosomal DNA restriction analysis (ARDRA) technique has been used to compare bacterial diversity and community composition from lake water samples, from agricultural soils, and from rhizosphere soil of diseased wheat.

The PCR-based random amplified polymorphic DNA (RAPD) molecular marker technique has been used to estimate the similarity of microbial communities in termite mounds and soils and in rhizosphere soil of maize. It has been illustrated that PCR-based molecular techniques can be used to investigate microbial communities in soil and to provide information on changes in the soil microbial communities that occur with different agricultural treatments and disease. These techniques need to be evaluated for the potential to characterise microbial communities in sugarcane soils.

The objectives of this study are to increase our basic understanding of microbial communities in sugarcane soils, to determine if there are differences in microbial communities associated with sugarcane roots in soil with and without a sugarcane cropping history, and to provide information on possible changes in the soil microbial communities that occur with sugarcane monoculture that may contribute to sugarcane yield decline.

In Louisiana, sugarcane (inter-specific hybrids of *Saccharum*) is grown largely under long-term monoculture production with successive annual harvests obtained from an initial vegetative planting of stalks. Therefore, sugarcane cultivars are clones composed of genetically identical plants. Buds on planted stalk sections germinate and grow to produce the first "plant cane" crop. Following the first harvest of the stalks, buds on the stubble remaining in the soil can germinate to produce a ratoon crop. In Louisiana, sugarcane plants are typically in the field continuously for 3-5 years. The stubble is then plowed out during the spring, and following a 4-6 month plant-free fallow period, the crop is replanted during late summer. Monoculture can lead to poor root health and growth constraints in many crops, including sugarcane.

A complex problem termed "yield decline" was defined by Garside, *et. al.* as "the loss of productive capacity of soils under sugarcane monoculture." Previous studies have provided evidence of yield decline in sugarcane. Yield responses to fumigation of 20% or greater have been obtained. Magarey demonstrated positive growth responses to soil solarisation and soil incorporation of high rates of broad-spectrum dithiocarbamate fungicides. Increased plant growth in soils not previously cropped to sugarcane provides additional evidence of yield decline. Research has been conducted in Louisiana on the effects of soilborne pathogens, such as *Pythium* and nematodes on sugarcane.

In greenhouse experiments, soil incorporation of Mancozeb has improved root health and plant growth, and organic material soil amendments reduced root rot severity and provided increases in plant growth not correlated with root colonisation by *Pythium*. However, the effects of the total soil microbial community on sugarcane growth in Louisiana are not well understood. One approach to study the total soil microbial communities associated with sugarcane roots growing in soils with a long-term sugarcane cropping history (SCH) and no recent sugarcane cropping history (NSCH) would be to compare the quantitative and qualitative aspects of culturable microorganisms.

It has been determined that only 0.01% to 10% of microorganisms in soil could be isolated and grown in culture. Therefore, culturable microorganisms represent only a small portion of the total community. However, as long as researchers remain aware of its limitations, selective media methods can be used to identify and quantify culturable species as a means to characterise and compare soil microbial communities. Traditionally, culture methods involve soil dilution plating on artificial media.

Medium components can be altered to selectively isolate and grow groups of microorganisms associated with plant roots. Culturable bacterial and fungal populations can be characterised to compare microbial communities. In addition, populations of selected organisms that can affect plant growth can be compared. *Pseudomonas* species found in association with the rhizosphere can play either a deleterious or beneficial role in plant growth and crop yields.

Rouatt and Katznelson reported that there were high numbers of Pseudomonas spp. in the rhizosphere soil of six different crop plants tested, including red clover, flax, oats, maize, barley and wheat. They suggested that their fast rate of growth, ability to produce acidic substances, and production of fluorescent pigments that were inhibitory to other organisms gave them a competitive advantage. *Pseudomonas* spp. have played beneficial roles as biocontrol agents of rhizosphere associated phytopathogens, including Fusarium wilts on tomato Rhizoctonia damping-off on cucumber roots and Gaeumannomyces graminis take-all of wheat. Some Pseudomonas spp. are pathogenic to lucerne, oats, and wheat.

Nehl, *et. al.* found that *Pseudomonas* spp., along with other rhizobacteria, fluctuate from being deleterious to plant growth promoting, depending on environmental conditions, host genotype and mycorrhizal status. It has been reported that phytopathogenic Pseudomonas syringae strains produce siderophores, which complex with ferric iron to provide themselves with iron under iron limited conditions and thereby gain a competitive advantage over other microorganisms. In contrast, Kloepper, *et. al.*, suggested that competition for iron is involved in soils that suppress plant diseases. In these soils, Pseudomonas spp. produce extracellular siderophores which may place phytopathogens at a disadvantage.

Actinomycetes produce antibiotics that are antagonistic to a number of different microorganisms, including some plant pathogens; however, some actinomycetes are phytopathogens themselves. Therefore, this group of organisms would be of potential interest in evaluating potential differences in soil microbial communities that might affect sugarcane growth. Phytopathogenic fungi have been extensively studied for many important crops, including sugarcane. Other fungi can suppress plant disease. Therefore, comparison of the total numbers and types of fungi associated with sugarcane roots in soils with and without a recent sugarcane cropping history would be desirable.

Fusarium spp. are abundant in agricultural soils, and many crops are damaged by Fusarium wilts. It has been demonstrated that *Fusarium* spp. can be pathogenic on sugarcane, so an evaluation of Fusarium populations could be of interest. The Biolog™ bacterial identification system was adapted to compare sole carbon substrate utilisation profiles (SCSUP) of microbial communities. SCSUP has been used to study the potential functional diversity of the culturable component of soil microbial communities associated with roots of hybrid larch, Sitka spruce, wheat, and grass species.

Biolog GN and GP MicroPlates™ contain 95 different carbon substrates and one control well containing no carbon substrate. In the Biolog microplate, a utilisation pattern appears as a colourless tetrazolium redox dye is reduced to a coloured formazan product when microorganisms utilise the substrates. For community comparisons, the SCSUP data is collected from the microplate and can be analysed by a multivariate analysis technique, such as principal component analysis (PCA).

The SCSUP data also can be used to calculate different measures of microbial community functional diversity, such as substrate richness (the number of utilised carbon sources), substrate evenness (the equitability of activities across all substrates), substrate diversity (encompasses both substrate richness and substrate evenness) (Shannon index), and diversity and evenness. However, SCSUP has limitations. The profiles only represent the activities of members of the community that grow rapidly under the conditions in the Biolog plates, and the ratio of numbers of replicates to the number of variables used may not be large enough to yield statistically stable results. However, Legendre and Legendre stated that the first few eigenvectors, those typically used in PCA, are little affected when the replicate-to-variable ratio is low. An alternative microplate designed for microbial community analysis the EcoPlate™ system, is composed of 31 carbon substrates useful for community analysis that are replicated three times on a single plate.

Three study sites were selected with SCH and NSCH fields either adjacent or closely situated to minimise variation in soil type, climate, and farm management. All sites were planted with the sugarcane cultivar LCP 85-384. All of the fields without a recent sugarcane cropping history were in the first season of sugarcane production. In contrast, all of the long-term sugarcane fields were in monoculture production for 15 or more years.

The first site was on a sugarcane farm in St. Charles Parish, Louisiana. The NSCH field was converted from a mixed-species pasture. The NSCH field plants were in the first year of planting, and the directly adjacent SCH field plants were in the first ratoon crop. The samples were collected in May, 2001. The second and third sites were on sugarcane farms in Iberia Parish, Louisiana. The NSCH field was converted from mixed hardwood forest at Site 2 and from a mixed-species pasture at Site 3.

At Site 2, the NSCH and SCH fields were on opposite sides of a waterway and separated by approximately 200 m. At Site 3, the NSCH and SCH fields were directly adjacent to each other. Samples were collected at Site 2 in June, 2002 and at Site 3 in August, 2002. At all sites, 25 samples were collected in a Z pattern from each field. Roots and soil were removed from a zone approximately 2-25 cm below plants, and roots were shaken so that only tightly adhering soil would be collected for rhizosphere soil samples.

The roots and adhering soil from each sample were placed in individual, sterile whirl-pak bags. The samples were placed in an ice chest with cold packs for transport to the lab. Non-rhizosphere soil also was collected. Rhizosphere soil samples from all sites were used for comparisons of culturable microorganisms on general and selective media and sole carbon source utilisation profiles. At Site 1, a composite of the Twenty five bulk soil samples was used for sole carbon source utilisation profile and selective media analyses. Thoroughly mixed composites of bulk samples from both fields at each site were used for fertility profile and particle size analyses. Bulk soil samples were dried at 105°C for 24 hr. to measure gravimetric soil moisture content.

Soil samples were stored at 4°C overnight before initiating experiments. Fertility profile analyses of NSCH and SCH fields included the following parameters: pH, content in parts per million (ppm) of salts, phosphorus (P), potassium (K), calcium (Ca), magnesium (Mg), iron (Fe), manganese (Mn), sulphur (S), zinc (Zn), copper (Cu), and content percentage of organic matter (OM). Content percentages of sand, silt, and clay from the particle size analyses were used to designate soil textures of the NSCH and SCH fields.

Preliminary research was performed to discover the optimal growing conditions for culturable organisms in microbial communities from sugarcane field soils. Various inoculating solutions were evaluated, including sterile, distilled water, Winogradsky salt solution pH 7.0, phosphate buffer solution, and calcium chloride solution. Various dilutions and volumes were evaluated. Types of media used to characterise the microbial community were 10% tryptic soy agar (TSA) for general bacterial colonies, two media selective for Pseudomonas species: King's B (KB) and Pseudomonas agar (PA), actinomycete isolation agar (AA) (Difco) selective for actinomycete colonies, and acidified potato dextrose agar (APDA) (PDA (Sigma) plus 2-ml of 25% lactic acid per liter) selective for fungal colonies.

Plate counts were expressed as colony forming units (cfu) per gram of dried soil. TSA, PA, and KB plates were placed under ultra-violet light to visualise and count fluorescent bacterial colonies. The means of five replicate plates were calculated. Data were $\log_{10}$ transformed before analysis with the general linear model (GLM) and least significant difference (LSD) statistical methods. The preliminary results revealed that soil serially diluted in 0.01 M $CaCl_2$ solution and 0.1 ml of inoculum per media plate was the optimal inoculation solution and volume for plating. The optimal dilutions to obtain countable colonies ranged from 10-2 to 10-4. The two Pseudomonas selective media were comparable to each other in colony producing efficiency. For convenience, PA was used instead of King's B.

For Site 1, air-dried soil samples were used for all media. For Sites 2 and 3, fresh or wet weight soil samples were used for all media except AA. Dried soil samples were used in addition to fresh with TSA and PA for comparison. Soil samples were aseptically air-dried for 2 wk as a pretreatment to inhibit growth of fungal species for actinomycetes cultivation. Roots were shaken vigorously and rubbed against each other inside the whirl-pak bags to collect rhizosphere soil.

The loose soil was collected from each of twenty five bags and combined in another sterile bag to create a composite sample for each field. Five replicate subsamples were then collected from the composite sample. Fresh soil was used within 24 hr from field collection. For future analysis, the remaining soil was poured into autoclaved, 60 ml glass jars with loosely sealed lids and placed in a covered box containing Drierite crystals and allowed to dry in the dark.

Selective media used for Sites 2 and 3 were the same as for Site 1, except that rosebengal (33 mg/L) streptomycin (30 μg/L) agar (RB) was used instead of APDA for general fungal colonies, chrome-azurol sulphate agar (CAS) selective for siderophore producing bacterial colonies, and Komada's agar (KO) selective for fusarial fungal colonies were added to the experiment. The chrome-azurol sulphate medium was modified from Schwyn and Neilands in that protease peptone #3 was used instead of casamino acid to cultivate organisms that would not grow on minimal medium.

Ten-fold serial dilutions of 1 ml of 0.01 M $CaCl_2$ were swirled into RB and KO media melted at 50°C and allowed to cool and harden. For Site 1, all plates were read after six days of incubation. For Sites 2 and 3, the CAS plates were read after 2 d of incubation, TSA and PA plates after 6 d, KO plates after 8 d, and AA plates after 12 d.

Evaluation of Microorganisms

To evaluate microorganisms associated with the interior of sugarcane roots from fields with and without a recent sugarcane cropping history, a modified method was used. From Sites 2 and 3, four roots from each of 25 replicate samples per field for a total of one hundred roots per sample were placed in a 1 L glass jar containing 250 ml of non-sterile dH_2O. This was replicated three times per field. To separate the rhizosphere soil from the roots, the roots were washed by shaking at 200 rpm for 20 min on a rotary shaker, and the soil/water suspension was discarded.

The wash was repeated six times. Then, 5 g of washed roots from each replication were measured and cut with alcohol flamed scissors. The measured roots were placed in sterile, 1 L glass jars containing 200 ml of 1% (v/v) final concentration of sodium hypochlorite (NaOClO) solution. The roots and bleach solution were shaken at 200 rpm for 10 min. The roots were then rinsed four times for 5 min each in 100 ml of sterile, 0.01 M $CaCl_2$ solution in a 250 ml sterile glass beaker. The rinse water from the last rinse was saved, and 0.1 ml was used to spread plate on TSA to check for surface contamination.

The roots were cut with flame-sterilised razor blades and tweezers, and the pieces were weighed into 250 mg amounts and transferred to sterile, 2 ml plastic centrifuge tubes with two tubes for each sample for a total of six tubes for each NSCH or SCH field. Each 2 ml tube contained a 5 mm glass bead and 1ml of 0.01 M $CaCl_2$. The tubes containing the roots were placed in a bead beater and shaken to macerate the root tissue for 30 sec. They were placed on ice for 2 min to cool and were shaken again for 30 sec. Then, the tubes were centrifuged at 7,000 g for 4-5 sec., and 1 ml of the supernatant was removed from tubes and mixed with 9 ml of $CaCl_2$ to make 10-fold serial dilutions of the samples. The dilutions (10^{-2} to 10^{-5}) were spread on 10% TSA. Colony forming units per ml of root extract on the plates were counted every two days for 12 d.

To gain qualitative information from the culture media experiments, bacterial colonies on TSA from all sites were isolated and stored at -20°C for future identification. For Site 1, colonies were not chosen at random. Colonies that differed by morphology from the NSCH and SCH field soil types were isolated and stored. However, from Sites 2 and 3, 40 colonies from each NSCH or SCH field soil were randomly picked and isolated from the culture plates and stored. After the final counts were taken, plates were placed over a 100 square grid (6 mm/square), and colonies from every other square were picked.

If the edge of a colony was touching other colonies, that one was skipped until a single colony surrounded by empty space was found. Three to four culture plates were required to obtain forty colonies for each field. The chosen colonies were then sub-cultured on TSA plates three times by dilution streaking to insure purity and placed on TSA slants. Preservation of the chosen colonies was conducted by modifying the method of Perkins. The colonies were washed off of the slants with 200 µl of a sterile skim milk solution pipetted into (13 x 100 mm) test tubes half-filled with 6-12 mesh, grade 40 silica gel.

The test tubes were placed in ice to decrease the high temperature produced from the dehydration of the solution and finally placed in storage at -20°C. Bacteria from the internal root colonising microorganism experiment for site 2 were isolated and stored at -20°C for future identification using the same storage method as for the selective media cultured isolates.

Determination of Sole Carbon Source Utilisation Profiles

Extraction of microbial cells for Site 1 soil samples was conducted by modifying the method of Grayston and Campbell. For rhizosphere samples, one root from each of twenty five replicate samples per field for a total of twenty five roots per sample was placed in a twiceautoclaved, 1L glass jar containing 250 ml of sterile phosphate buffer (0.01 M K_2HPO_4, 0.01 M KH_2PO_4, 0.15 M NaCl, and distilled H_2O, pH 7.0). This was replicated three times per field. The phosphate buffer and roots were shaken at 100 rpm for 1.5 h on a rotary shaker to separate the rhizosphere soil from the roots. The rhizosphere soil suspensions were poured into 250 ml sterile polypropylene bottles, and ten-fold serial dilutions (3 ml of solution into 27 ml of 0.01 M $CaCl_2$) were made from the suspensions.

Likewise, ten-fold serial dilutions of bulk soil suspension were made. For each soil sample, 30 ml of 10^{-3} dilution was placed in an autoclaved, 50 ml centrifuge tube and centrifuged at 750 g for 10 min at 4°C to pellet any soil particles that could interfere with optical density determination. Aliquots of 150 µl were then added to each of 96 wells of Biolog gram-negative (GN2) MicroPlates. There were three replicate plates per field. The plates were incubated at 25°C in the dark. Each well in a plate was measured for colour absorbance at 600 nm every 24 h for 5 d with a microplate reader. For NSCH and SCH soil samples from Sites 2 and 3, the inoculum was extracted and standardised by modified methods.

Four roots from each of twenty five sub-samples per field for a total of one hundered roots per sample were placed into each of five replicate twice-autoclaved, glass, 1 liter jars with 250 ml of phosphate buffer. The phosphate buffer and roots were shaken as for Site 1. Instead of using ten-fold serial dilutions, the solutions were centrifuged at 2,000 x g for 20 min at 4°C to pellet any soil. The supernatants were then poured into 250 ml polypropylene copolymer centrifuge bottles and centrifuged at 10,000 x g for 10 min at 4°C to collect bacterial pellets.

The supernatants were discarded, and the pellets were vortex-mixed in 25 ml of sterile, 0.85% sodium chloride. The five sub-samples were then combined to create a composite sample. Three replicate samples were then made from the composite sample. Any remaining soil particles were allowed to settle for 5 min, and 1 ml from each replicate was used to read optical density at 600 nm. To standardise each replicate, absorbance values were adjusted to 0.08-0.10 O.D. The adjusted samples were shaken for 24 hr at room temperature to allow time for the bacteria to utilise any extraneous carbon in the suspension.

Aliquots of 150 µl of the standardised samples were then added to each well in Biolog GN2™ plates. The plates were incubated at 28°C in the dark. Each well in the Biolog plates was then measured for absorbance as for Site 1. Preliminary research was performed with the Biolog plate absorbance data to determine the analysis method that would reveal the most useful information. The plates were analysed using the following data sets: means of individual well data of the three replicate plates for each day compared separately and means for individual wells over days 2, 3, and 4, for each plate. The fifth day data was not used because of excessive growth and fungal contamination.

The averaged data whether for single or multiple dates were dissimilar to visually determined results in single plates on single dates. Therefore, single day, individual plate data were analysed by principal component analysis (PCA) to compare SCSUP for each site. Data used for PCA were from the single day reading with the maximum amount of positive wells before contamination by fungi occurred. The data recorded at 72 hr were used for Site 1, and at 24 hr for Sites 2 and 3.

The first principal component was related to overall plate responses, and the second and third principal component were related to individual carbon substrates. A covariate matrix was used because the units of measurement were the same. Microbial communities from individual plate wells were preserved for future molecular profiling of samples from Sites 2 and 3. Microbial communities in Biolog plates were pipetted out of individual wells that differed between NSCH and SCH field soil types and stored at -20°C in 500 µl of sterile, dH_2O in 1.5 ml micro-centrifuge tubes.

Soil Fertility Profile and Particle Size Analyses

For all three sites, pH, organic matter, and fertility parameters were generally numerically higher in NSCH fields, except for Site 3 where pH, salt, and K were higher in SCH fields. The soil textures of each field at Sites 1 and 2 were silt loam. The soil textures of each field at Site 3 were different from each other and from Sites 1 and 2. At Site 3, the NSCH field soil texture was sandy clay loam and the SCH field was loam (Table 1).

For Site 1, total bacteria cfu/g of soil in bulk soil was higher in NSCH than SCH on TSA, PA, and AA media (Fig. 1). However, there was no difference between SCH and NSCH rhizosphere soils for total bacteria cfu/g of soil on TSA, PA, and AA media. On TSA, fluorescent bacteria were isolated from rhizosphere soil of NSCH but not SCH samples, with a mean cfu/g of soil of $4.0 \times 10^4 \pm 7.1 \times 10^3$ (standard error), and none were isolated from bulk soils. Fluorescent bacterial colonies were isolated from both NSCH and SCH samples on PA.

Fluorescent colonies were more numerous on PA plated with diluted rhizosphere soil from NSCH samples with a mean cfu/g of soil of $1.5 \times 10^6 \pm 3.2 \times 10^5$

Table 1. Soil fertility profiles and particle size analyses of experimental sugarcane fields.

Site	Cropping history	pH	Salts	P	K	Ca	Mg	Fe	Mn	S	Zn	Cu	OM	Sand	Silt	Clay
		Content (ppm)										Content (%)				
1	NSCH	6.6	336	232	274	2,958	617	84	14	9	6	6	2.1	12	73	15
1	SCH	6.1	112	207	174	1,606	325	64	14	5	2	3	1.5	10	68	22
2	NSCH	7.0	1,493	199	200	2,491	441	48	34	26	4	1	2.1	5	75	20
2	SCH	6.8	898	180	115	2,904	383	34	27	43	2	1	1.6	5	78	17
3	NSCH	5.8	297	99	158	1,776	325	114	36	18	3	2	3.0	52	15	33
3	SCH	6.1	328	78	219	1,241	189	65	32	17	1	1	1.8	44	31	25

compared to 1.7 x 10^4 ± 3.2 x 10^5 for SCH samples, and none were detected in bulk soil platings as with TSA. Each APDA plate produced fewer than five colonies; therefore, fungal cfu/g could not be determined and compared.

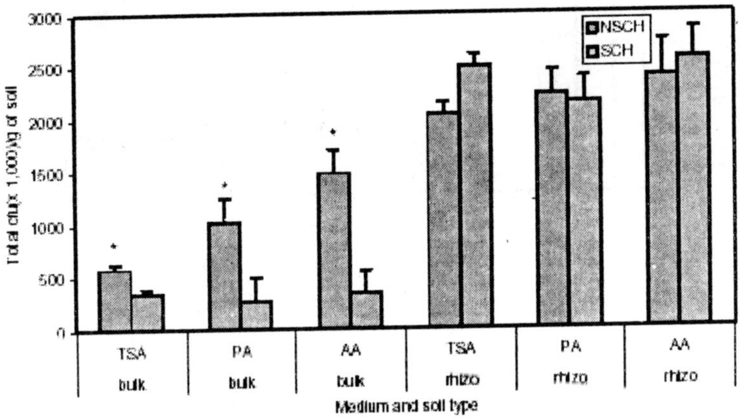

Figure 1. Total microbial cfu/g of soil on tryptic soy agar (TSA), Pseudomonas agar (PA), and actinomycetes agar (AA) from bulk and rhizosphere (rhizo) soils from fields with no recent sugarcane cropping history (NSCH) and long-term sugarcane cropping history (SCH) at Site 1.

For Site 1, cfu/g of soil of microbial organisms was compared on different media for bulk and rhizosphere soil samples from NSCH and SCH fields (Table 2). On TSA, PA, and AA, the total cfu/g of soil was higher in rhizosphere than bulk soil, except on AA from NSCH samples. For Site 2, only rhizosphere soil was used as inoculum for all media. Total cfu/g of soil was higher in SCH soil on TSA with fresh inoculum, while there was no difference between NSCH and SCH with the air-dried soil inoculum on TSA (Fig. 2). In contrast, there was no difference in total bacterial cfu/g of soil on PA with fresh soil inoculum; however, with air-dried soil inoculum, total cfu/g of soil was higher in NSCH soil.

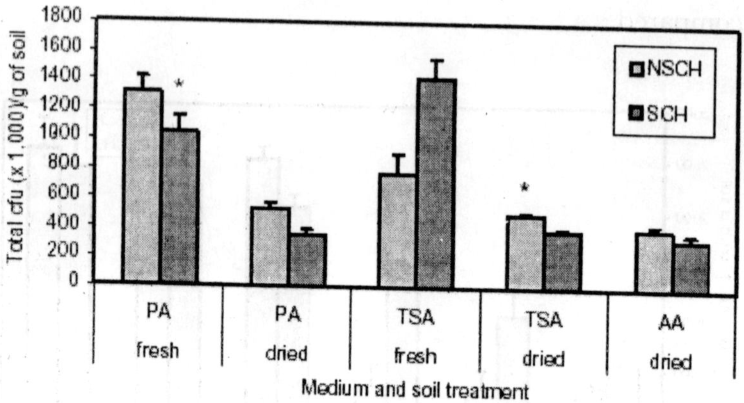

Figure 2. Total microbial cfu/g of soil on tryptic soy agar (TSA), Pseudomonas agar (PA), and actinomycetes agar (AA) using fresh and air-dried soil inocula from fields with no recent sugarcane cropping history (NSCH) and long-term sugarcane cropping history (SCH) at Site 2.

There was no difference in cfu/g of soil between the soil types on AA medium. Fluorescent bacterial colonies did not appear on TSA with fresh or air-dried soil inoculum in either SCH or NSCH soils. Fluorescent bacterial cfu/g of soil was higher in NSCH soil with air-dried but not fresh soil inoculum on PA. There was no difference in cfu/g of soil between SCH and NSCH soils for total or siderophore producing bacteria on CAS. Total fungal cfu/g of soil was higher in NSCH soil with fresh soil inoculum for fungi on RB. On KO medium, more fungi were detected in NSCH than SCH soil. The mean cfu/g of soil was $1.3 \times 10^3 \pm 5.3 \times 10^1$ in NSCH soil and $7.1 \times 10^1 \pm 5.3 \times 10^1$ in SCH soil.

For Site 3, only rhizosphere soil was used for inoculum for plating on all media. Total bacterial cfu/g of soil was higher in NSCH than SCH soil on TSA and AA with air-dried soil inoculum. There was no difference in total bacterial cfu/g of soil on TSA with fresh inoculum and on PA with fresh or air-dried soil inoculum. Fluorescent bacteria did not appear on TSA with fresh or air-dried soil inoculum from either field. For Site 3, there was no difference between NSCH and SCH in fluorescent bacteria cfu/g of soil on PA with air-dried or fresh soil inoculum. Total and siderophore producing bacteria cfu/g of soil were higher in SCH soil on CAS medium.

Table 2. Comparison of total bacterial cfu/g of soil in bulk and rhizosphere soil samples collected from no recent sugarcane cropping history (NSCH) and sugarcane cropping history (SCH) fields at Site 1.

	Field type and culture medium		
Soil type		*NSCH*	
	TSA	*PA*	*AA*
Rhizosphere	$2.0x10^6 \pm 9.0x10^4$	$2.2x10^6 \pm 1.8x10^5$	$2.4x10^6 \pm 4.3x10^5$
Bulk	$5.8x10^5 \pm 7.8x10^4$	$1.0x10^6 \pm 1.8x10^5$	$1.5x10^6 \pm 2.7x10^5$
Soil type		*SCH*	
	TSA	*PA*	*AA*
Rhizosphere	$2.5x10^6 \pm 8.6x10^4$	$2.2x10^6 \pm 2.0x10^5$	$2.5x10^6 \pm 8.7x10^4$
Bulk	$3.4x10^5 \pm 7.4x10^4$	$3.6x106 \pm 2.0x10^5$	$3.3x10^5 \pm 1.1x10^5$

There was no difference in total fungal cfu/g of soil detected on RB in either soil. On KO medium, the fungal mean cfu/g of soil was higher in NSCH ($1.0 \times 10^3 \pm 8.2 \times 10^1$) than SCH ($4.8 \times 10^2 \pm 9.2 \times 10^1$). A comparison of microbial cfu/g of soil detected on different media considered by experimental sites revealed differences among Sites 2 and 3 for NSCH and SCH field types. Site 1 was not compared because the samples were processed differently. The mean cfu/g of soil of fluorescent bacteria on PA, siderophore producing bacteria on CAS, and fungi on KO were not different between Sites 2 and 3 for NSCH soil, and fluorescent bacterial counts were not different in SCH. For all other media types, the numbers of microorganisms detected were higher at Site 3 for both field types.

For Site 2, the numbers of total bacterial cfu/ml of root extract from samples on 10% TSA medium were either greater than 300 or less than 30 with all dilutions. Therefore, the colonies were not countable for statistical evaluation. However, 40 colonies each from NSCH and SCH soils were collected at random from the TSA plates and saved for future identification and comparison. For Site 3, the detection of bacterial colonies from all samples on TSA plates inoculated with the last rinse solution indicated that microorganisms on the root exterior may not have been completely eliminated. Therefore, no colonies were saved from the Site 3 samples.

For all sites, field types, rhizosphere, and root interiors, the most common type of bacterial colony on all media was white to off-white, circular shaped, and 3-4 mm in diameter (dia.) after 6 d of growth. For Site 1, colonies were not chosen at random. Instead, ones that differed in morphology were isolated from the NSCH and SCH field soil types and stored for future identification. In general, yellow colony types appeared following plating of the rhizosphere soil but not of the bulk soil, regardless of NSCH and SCH soil type.

For Site 2, NSCH rhizosphere soil plating yielded 25 different colony types among the 40 randomly isolated colonies. Most of the different colony types represented single colonies, but four were white and 5 mm in dia., another four were white and 1 mm in dia., and five were yellow and 3 mm in dia. Plating SCH soil yielded 22 different colony types among the 40 isolated colonies. Again, most of the different colony types were represented by single colonies. Three were yellow and 4 mm in dia., and four were white and 1 mm in dia.

For Site 3, plating of NSCH rhizosphere soil yielded 17 different colony types among the 40 randomly isolated colonies. Three were yellow and 5 mm in dia, and nine were yellow and 4 mm in dia. Sixteen different colony types were isolated from SCH soil. Most types were represented by a single colony, and three were yellow and 4 mm in dia. For Site 2, plating of NSCH root extracts yielded 10 different colony types among the 40 randomly isolated colonies. Two were yellow and 3 mm in dia. Eight different colony types were isolated from SCH root extracts. Most types were represented by a single colony, and three were white with leathery surfaces and 3 mm in dia.

Comparison of NSCH and SCH Sole Carbon Source Utilisation Profiles

At Site 1, rhizosphere microbial communities in soil with or without a recent sugarcane cropping history could be distinguished by PCA of sole carbon source utilisation profiles. However, there was high variability among the NSCH replicates, and one replicate showed similiarity to the SCH samples. The PCA of rhizosphere soil samples indicated that the first (overall plate reaction) and second (individual well reactions) principal components explained 55% of the standardised variance, and the first three principal components explained 72%.

Also at Site 1, SCH and NSCH bulk soil microbial communities could be distinguished by PCA of sole carbon source utilisation profiles. However, one of the SCH replicates was different than the other two. The PCA of bulk soil samples indicated that the first and second principal components explained 64% of the standardised variance, and the first three principal components explained 84%. Only twelve substrates were utilised overall in the bulk soil samples at Site 1. Therefore, the absorbance values of the other 83 substrates were discarded to rid the PCA of redundant data.

For bulk soils, the carbon substrates utilised in SCH soil but not NSCH were Dmannose, D-galacturonic acid, D-saccharic acid and D, L-carnitine, and substrates utilised in NSCH but not SCH were D-gluconic acid, β-hydroxybutyric acid, L-histidine, and L-leucine, L-pyroglutamic acid, and 2-amino ethanol. Microbial communities from both types of bulk soil utilised the carbon substrates D,L-lactic acid and L-glutamic acid.

At Site 1, SCH and NSCH bulk and rhizosphere soil microbial communities could not be clearly distinguished by PCA of sole carbon source utilisation profiles. One of the replicate SCH bulk soil was separated from the other two SCH bulk samples by principal component 1. The replicates of SCH rhizosphere soil did not cluster tightly, and the NSCH rhizosphere replicates did not cluster at all. The GN2 plate data PCA indicated that the first and second principal components explained 62% of the standardised variance, and the first three principal components explained 71%.

At Site 2, SCH and NSCH rhizosphere soil microbial communities could be distinguished by PCA of sole carbon source utilisation profiles. However, one replicate plate from each of the soil types was separated from the other two of its type by principal component 2. The GN2 plate data PCA indicated that the first and second principal components explained 90% of the standardised variance, and the first three principal components explained 94%. For Site 2 rhizosphere soil, six carbon substrates were utilised by the rhizosphere community in SCH samples but not NSCH samples, including glycogen, tween 40, propionic acid, succinamic acid, 2-aminoethanol, and glycerol.

Three substrates were utilised in rhizosphere NSCH samples but not SCH samples: gentiobiose, acetic acid, and malonic acid. Microbial communities in both types of rhizosphere samples utilised numerous carbon substrates in common, including N-acetyl-D- glucosamine, adonitol, L-arabinose, cellobiose, D-fructose, D-galactose, α-D-glucose, maltose, D-mannitol, Dmannose, D-melibiose, β-methyl-D-glucoside, D-psicose, D-raffinose, L-rhamnose, sucrose, D-trehalose, methyl pyruvate, mono-methyl succinate, cis-aconitic acid, citric acid, D-galacturonic acid, D-gluconic acid, D-glucuronic acid, p-hydroxyphenylacetic acid, D, L-lactic acid, quinic acid, D-saccharic acid, succinic acid, bromosuccinic acid, alaninamide, D-alanine, L-alanyl-glycine, L-aspartic acid, L-glutamic acid, L-leucine, Lornithine, L-phenylalanine, L-proline, D-serine, L-serine, D, L-carnitine, urocanic acid, inosine, uridine, thymidine, D, L-α-glycerol phosphate, and glucose-6-phosphate.

At Site 3, SCH and NSCH rhizosphere soil microbial communities were again separated by PCA of sole carbon source utilisation profiles. Variability was detected in one SCH sample for overall plate reaction (principal component one) and among the NSCH sample replicates for some substrates utilised (principal component two). The GN2 plate data PCA indicated that the first and second principal components explained 72% of the standardised variance, and the first three principal components explained 90%.

For Site 3 rhizosphere soil, five carbon substrates were utilised in rhizosphere SCH samples but not NSCH samples: glycogen, p-hydroxyphenylacetic acid, succinamic acid, D-serine, and phenylethylamine, while four substrates were utilised in rhizosphere NSCH samples but not SCH samples, including D-mannitol, D-glucuronic acid, β-hydroxybutyric acid, and glucose-6-phosphate.

Both types of rhizosphere samples utilised numerous carbon substrates in common, including dextrin, tween 80, N-acetyl-D-glucosamine, sucrose, D-trehalose, methyl pyruvate, cis-aconitic acid, citric acid, D-galacturonic acid lactone, D-gluconic acid, α-ketoglutaric acid, D, L-lactic acid, quinic acid, D-saccharic acid, succinic acid, bromosuccinic acid, L-asparagine, L-aspartic acid, L-glutamic acid, L-hisitdine, hydroxy-L-proline, L-proline, L-pyroglutamic acid, L-serine, γ-aminobutryic acid, urocanic acid, inosine, putrescine, glycerol, and glucose-6-phosphate.

Comparing SCH rhizosphere soil at all three sites, the soil microbial communities were separated by PCA of sole carbon source utilisation profiles. Variability was detected in one Site 1 sample for overall plate reaction (principal component one). The GN2 plate data PCA indicated that the first and second principal components explained 54% of the standardised variance, and the first three principal components explained 88%.

Comparing both NSCH and SCH rhizosphere soil at all three sites, the soil microbial communities for the different sites were separated by PCA of sole carbon source utilisation profiles rather than by NSCH and SCH soil types. The GN2 plate data PCA indicated that the first and second principal components explained 59% of the standardised variance, and the first three principal components explained 76%.

6

Soil Health

Healthy soils are essential for the integrity of terrestrial ecosystems to remain intact or to recover from disturbances, such as drought, climate change, pest infestation, pollution, and human exploitation including agriculture. Protection of soil is therefore of high priority and a thorough understanding of ecosystem processes is a critical factor in assuring that soil remains healthy.

To manage and maintain soil in a sustainable fashion, the definition of soil health must be broad enough to encompass the many functions of soil, *e.g.* environmental filter, plant growth and water regulation. Definitions of air and water quality standards have existed for a long time, while a similar definition does not exist for soil. There is, however, little if any parallel between air or water quality and soil health. Air and water quality standards are usually based on maximum allowable concentration of materials hazardous to human health. Soil health is the net result of on-going conservation and degradation processes, depending highly on the biological component of the soil ecosystem, and influences plant health, environmental health, food safety and quality.

Historically, the term soil quality described the status of soil as related to agricultural productivity or fertility. In the 1990s, it was proposed that soil quality was not limited to soil productivity but instead expanded to encompass interactions with the surrounding environment, including the implications for human and animal health. In the mid-1990s, the term soil health was introduced.

Soil is a finite and non-renewable resource because regeneration of soil through chemical and biological weathering of underlying rock requires geological time. Deterioration of soil, and thereby soil health, is of concern for human, animal and plant health because air, groundwater and surface water consumed by humans can be adversely affected by mismanaged and contaminated soil. As such, deteriorating soil health and the benefits of soil management has become of political concern.

A healthy soil functions to buffer nutrients as well as contaminants and other solutes via sorption to or incorporation into clay particles and organic materials. The soil itself thus serves as an environmental filter for removing undesirable solid and gaseous constituents from air and water. The extent to which a soil immobilises or chemically alters substances that are toxic, thus effectively detoxifying them, reflects the degree of soil health in the sense that humans or other biological components of the system are protected from harm. Soil is dominated by a solid phase consisting of particles of different size surrounded by water and gases, the amount and composition of which fluctuate markedly in time and space.

Water is normally discontinuous, except when the soil is water saturated. The pore space without water is filled with air and other gases and volatiles. There is continual interchange of molecules and ions between solid, liquid and gaseous phases which are mediated by physical, chemical and biological processes. These processes represent a unique balance between physical, chemical and biological components. Maintaining this balance is of great importance to soil health. The biological activity in soil is largely concentrated in the topsoil, the depth of which may vary from a few to 30 cm.

In topsoil, the biological components occupy a tiny fraction (<0.5%) of the total soil volume and make up less than 10% of the total organic matter in soil. These biological components consist mainly of soil organisms, especially microorganisms. Despite their small volume in soil, microorganisms are key players in the cycling of nitrogen, sulphur, and phosphorus, and the decomposition of organic residues. Thereby, they affect nutrient and carbon cycling on a global scale. That is, the energy input into the soil ecosystems is derived from the microbial decomposition of dead plant and animal organic matter.

The organic residues are, in this way, converted to biomass or mineralised to CO_2, H_2O, mineral nitrogen, phosphorus, and other nutrients. Mineral nutrients immobilised in microbial biomass are subsequently released when microbes are grazed by microbivores such as protozoa and nematodes. Microorganisms are further associated with the transformation and degradation of waste materials and synthetic organic compounds. In addition to the effect on nutrient cycling, microorganisms also affect the physical properties of soil. Production of extra-cellular polysaccharides and other cellular debris by microorganisms help in maintaining soil structure, as these materials function as cementing agents that stabilise soil aggregates. Thereby, they also affect water holding capacity, infiltration rate, crusting, erodibility, and susceptibility to compaction.

Role of Microorganisms

Microorganisms possess the ability to give an integrated measure of soil health, an aspect that cannot be obtained with physical/chemical measures and/or analyses of diversity of higher organisms. Microorganisms respond quickly to changes, hence they rapidly adapt to environmental conditions. The microorganisms that are best adapted will be the ones that flourish. This adaptation potentially allows microbial analyses to be discriminating in soil health assessment, and changes in microbial populations and activities may therefore function as an excellent indicator of change in soil health. Microorganisms also respond quickly to environmental stress compared to higher organisms, as they have intimate relations with their surroundings due to their high surface to volume ratio.

In some instances, changes in microbial populations or activity can precede detectable changes in soil physical and chemical properties, thereby providing an early sign of soil improvement or an early warning of soil degradation. An example is the turnover rate of the microbial biomass. This is much faster, *e.g.* 1-5 years, than the turnover of total soil organic matter. Observations in the Dutch Soil Monitoring Programme have shown that most microbial indicators indeed have discriminating power relative to different soil treatments. This has also been shown for microbial biomass and basal respiration at a regional scale in the USA. The bioavailability of chemicals, *e.g.* heavy metals or pesticides, is also an important issue of soil health because of its connection with microbial activities. The impact of such chemicals on soil health is dependent on microbial activities.

For example, the concentration of heavy metals in soil will not change over small time periods, but their bioavailability may. It has thus been shown that the bioavailability of poly-aromatic hydrocarbons was lower in autumn compared to early spring due to a higher microbial activity after the growing season. Therefore, the total content of chemicals in soil is not a reliable indicator of its bioavailability and thereby soil health. Instead, bioavailability has to be measured in relation to bioassays and specific microbial processes.

Evaluation of Soil Health

Evaluation of soil health should be considered relative to the many different land uses, *e.g.* agriculture, forestry, urbanisation, recreation, and preservation. The objective for evaluating soil health in an *e.g.* agricultural ecosystem may, consequently, be different from objectives used for assessing urban or natural ecosystems. Thus, in agriculture, soil may be managed to maximise production without adverse environmental effects, whereas in a natural ecosystem, soil may be managed by a set of baseline values against which future changes in the system may be compared.

A framework for soil health evaluation is critical for the development of a useful monitoring programme covering the different functions and land-uses and it must identify priorities and relevant indicators relating to policy-relevant end points. An overall framework for soil health evaluation in Europe has recently been proposed by the European Environmental Agency (EEA), though it has yet to be implemented.

The main objective of this is to provide policy-makers with relevant environmental parameters based on reliable and comparable data related to soil and to facilitate comprehensive reporting on the state of soils in Europe. It also provides consistent measuring and assessment at any site, from handling of soil samples to the evaluation and storage of data. A similar framework has been used for arable soils in Sweden. Definition of policy-relevant end points is very important as monitoring programmes are developed. End points should be pragmatic in the sense of providing logical categories for regulatory decisions and they should be integrated for indicators that are ecologically related.

After reviewing environmental monitoring programmes, it is clear that end points for soil health need to be clearly specified and then used as guidance in the identification of indicators.

Selection of Microbial Indicators

Instead, a minimum data set (MDS) consisting of the smallest number of indicators needed to address the specific end point should be defined. Besides microbial indicators, a MDS for soil health monitoring should also include physical, chemical and biological indicators. A MDS is based on the objective of the monitoring programme and may very well be different for different end points.

Furthermore, the optimal MDS vary for different soil types and regions, since indicators vary due to climate, topography, parent material, vegetation and land use practices. Representatives of both inherent and dynamic soil components should be included in a MDS. Inherent soil properties are determined by the basic soil forming factors, including the geological material, climate, time, topography and vegetation. Dynamic soil properties are based on biological activity and include microbial indicators.

In the following, only microbial indicators will be dealt with as a part of a MDS. Full soil profile descriptions together with data of a range of physical and chemical properties are available for Danish soils. Generally, indicators of a MDS should be selected on the basis of their ease of measurements, reproducibility, and their sensitivity towards key variables controlling soil health. Each microbial indicator, however, represents slightly different aspects of soil health and has its advantages and disadvantages.

The selection of indicators should be broad enough to give policymakers and/or the general public an overview of the state of the environment, while detailed indicators are needed to better understand underlying trends. For microbial indicators, the overview will be accomplished by measurements at the ecosystem level (*e.g.* processes), which have been proposed to offer the best possibilities for rapidly assessing changes in soil health. Resulting data will allow decisions to be made at the community (*e.g.* biomass and biodiversity) or population (*e.g.* species or functions) levels and allow these detailed studies to be planned more precisely.

Ranking of the indicators according to applicability, economy, ease of interpretation, development needs, sensitivity etc. has also been proposed as a way to select the optimal indicator of soil health. Indicators can also be selected on basis of whether they are laboratory (in vitro) or field (in situ) measurements. In vitro measurements may involve incubation of a soil sample in the laboratory under standardised conditions and thus provide an estimate of the potential of the soil. Interpretation of in vitro measurements in relation to soil health can be difficult, since the results depend on the incubation conditions, which may not be comparable to field conditions. Examples of in vitro measurements are soil respiration, CFI/CFE, SIR, N-mineralisation, nitrification, denitrification, MPN and other growth-based methods.

In situ measurements are based either on direct measurements in the field or fixed samples analysed in the laboratory. They give a "snap-shot" measurement of the conditions in the soil. In situ measurements, however, are often very sensitive to spatial and temporal variation and this may over-ride the variability in soil health status. Examples of in situ measurements are gas emissions, PLFA, organic matter decomposition, thymidine and leucine incorporation, short-term enzyme assays and most molecular methods.

Integration of more indicators into one single method may be a way to reduce the number of indicators. At present, only few methods provide such integrated information. The phospholipid fatty acid (PLFA) analysis provides information about soil microbial biomass, fungal-bacterial ratio, biodiversity and occurrence of key species in one analysis. Substrate induced respiration (SIR) provides measurement of basal respiration and soil biomass. Finally, the carbon utilisation pattern (BIOLOG™) provides a profile of the microbial community and information on potential metabolic capacity, which together comprise functional diversity.

It has recently been noted that measurements relating to early changes in organic matter and biological and microbial attributes are particularly underrepresented in existing soil monitoring networks world-wide, although these are emerging areas of interest to the scientific community. Experience with the use of microbial indicators in soil monitoring is available in some European countries, where the most commonly used indicators are microbial biomass and soil respiration.

Classification of Microbial Indicators

Indicators of Biodiversity

Microbial indicators of soil health encompass a diverse set of microbial measurements due to the multi-functional properties of microbial communities in the soil ecosystem. In the present catalogue, bacteria, fungi and protozoa indicators are considered. They are grouped according to the different soil health parameters of the ecosystem, that is biodiversity, carbon cycling, nitrogen cycling, biomass, microbial activity, key species and bioavailability.

The indicators relate to the ecosystem (*e.g.* processes), community (*e.g.* biomass and biodiversity) or population (*e.g.* species or functions) levels and this relationship is noted together with relations to policy-relevant end point. The catalogue presents a comprehensive list of microbial indicators, some of them are ready-to-use in a monitoring programme, while others have to be developed for that purpose. Only a limited number of indicators will be implemented in a monitoring programme and a selection of indicators is necessary.

Information about microbial community structure and diversity has been noted as important for understanding the relationship between environmental factors and ecosystem functions. Microbial diversity measurements have thus been recommended in soil health monitoring programmes and represents measurements at the community level. The diversity of a community is expressed as the species richness and the relative contribution each species makes to the total number of organisms present. Diversity of a microbial community is often described by the Shannon-Weaver index (H′). The number of species has traditionally been determined by taxonomic classification studies, but as these are suboptimal for microorganisms, molecular and biochemical techniques of estimating abundance and number of each species must be applied.

The benefit of a high genetic diversity is currently under debate because it is not always correlated to functional diversity. Furthermore, the correlation between soil health and biodiversity is not completely understood, although a medium to high diversity is generally considered to indicate a good soil health.

Genetic microbial diversity

The genetic resources present in the environment are the basis of all actual and potential functions. The genetic diversity of soil microorganisms is an indicator of the genetic resource. Methods for determination of the genetic microbial diversity include several molecular methods of which a few may be implemented into a soil monitoring programme.

Bacterial diversity: Genetic diversity of bacteria is most commonly studied by diversity of the 16S rDNA genes, which occur in all bacteria and which show variation in base composition among species. 16S rDNA genes are thus used for phylogenetic affiliation of Eubacteria and Archaea and large databases exist on sequences of 16S rDNA. It consists of variable and conserved regions, and this has facilitated the design of primers in the conserved regions for targeting the majority of members of defined groups of bacteria. Two methods have been developed to examine the diversity of 16S rDNA sequences in total DNA extracted from soil microbial communities, namely PCR-DGGE and T-RFLP. Denaturing Gradient Gel Electrophoresis (PCR-DGGE) and Temperature Gradient Gel Electrophoresis (PCR-TGGE) are based on variation in base composition and secondary structure of fragments of the 16S rDNA molecule.

By PCR with primers principally targeting all eucaryotes or selected subgroups, a fragment of 16S rDNA of known size can be amplified. Following PCR, the products are separated by gel electrophoresis. By PCR-DGGE the gel itself contains a chemical-denaturing gradient, making the fragments denature along the gradient according to their base composition. By PCR-TGGE a temperature gradient is created across the gel, resulting in the same type of denaturation. The number and position of fragments reflect the dominating bacteria in the community. For the PCR-DGGE and PCR-TGGE methods, the low resolution of gel electrophoresis compared to the high diversity of bacterial communities can be a problem.

Soil communities may easily contain several hundred bacterial strains, while the resolution of more than 20-50 bands on a gel is difficult. For a visible band on the gel, a bacterial species has to comprise approx. 1% of the entire population, of course dependent upon many practical circumstances. Sequencing and identification of the visible bands on the gel following PCR-DGGE or PCR-TGGE may further improve the resolution of the important players of a microbial community. PCR-DGGE has recently been implemented in the Dutch Soil Monitoring Programme. Results from the first year showed that the mean number of DNA bands was found to be about 50 independent of the season. Furthermore, no sig-nificant difference was found between a dairy farm on clay soil and a horticultural farm on sand.

Terminal Restriction Fragment Length Polymorphism (T-RFLP) is an alternative method for examining diversity of 16S rDNA sequences of microbial communities. It is also based on PCR amplification of 16S rDNA with specific primers. The primers are labelled with a fluorescent tag at the terminus resulting in labelled PCR-products. The products are cut with several restriction enzymes, one at a time, which result in labelled fragments that can be separated according to their size on agarose gels. As the PCR products are labelled at the terminus, only restriction enzyme fragments containing either of the terminal ends of the PCR product will be detected.

The digested PCR products are subsequently loaded on a sequencer. The output includes fragment size and quantity. Recently, the potential of the T-T-RLFP method to discriminate soil bacterial communities in cultivated and non-cultivated soils has been demonstrated. As data accumulate and become accessible the method will allow comparison between different soils analysed in different labs. The method, however, requires delicate and expensive instruments along with very pure DNA.

Fungal diversity: The classical method for estimating the fungal diversity of soil has been number and morphology of fruiting bodies. However, the majority of fungi in soil are present either as resting stages (spores) or mycelium. Both spores and mycelium can be isolated from soil, but if a fruiting body is not formed, identification of the organisms is difficult at best, and generally impossible. Further, the isolation step may be selective to specific fungal groups, *e.g.* the fast growing ones.

Molecular methods based on 18S rDNA provide tools that can overcome these problems. However, a major limitation is the limited number of fungal nucleic acid sequences presently available in databases. Diversity measurements within the fungal community in soil can also be measured by PCR-DGGE and PCRTGGE. The methodologies are described above in relation to bacterial genetic diversity.

Protozoan diversity: Protozoae is a phylum of single cell eucaryotic organisms and as such may resemble and better represent higher organisms than prokaryotes. Protozoa are a paraphyletic group primarily consisting of naked amoebae, testate amoebae, ciliates and heterotrophic flagellates. Protozoa are very abundant in soil, like bacteria, and exist in very diverse and harsh environments. They also resemble bacteria in that they are important for soil health and fertility, react quickly to environmental changes, are ubiquitous and do not easily move around in soil.

Protozoa form an essential part of all soil ecosystems and have been proposed as early warning indicators. Protozoan bioassays, for example, have been used in the United Kingdom as a discriminating indicator of heavy metal contamination in soil amended with sewage sludge. Determination of the diversity of protozoa is normally carried out by taxonomic affiliation to species, groups or families based on morphological features. This method is very time consuming, requires specialists and is further complicated by the incomplete taxonomic description of protozoa.

Alternatively, protozoan diversity can be determined by molecular methods. The diversity of protozoa has been characterised by PCR-DGGE targeting an 18S rDNA fragment. The PCR-DGGE method is described above in relation to bacterial genetic diversity. Recently, a method of PCR-DGGE specific to Kinetoplastida, a monophyletic group of protozoa, has been published revealing a relatively high diversity in freshwater sediments. As protozoa form a paraphyletic group, specific primers for the various important protozoan groups have to be designed. The technique is being developed in these years and further development is necessary before implementing into a monitoring programme.

Functional diversity of microbial population

The diversity of functions within a microbial population is important for the multiple functions of a soil. The functional diversity of microbial communities has been found to be very sensitive to environmental changes. However, the methods used mainly indicate the potential in vitro functionality. Functional diversity of microbial populations in soil may be determined by either expression of different enzymes or diversity of nucleic acids (mRNA, rRNA) within cells, the latter also reflecting the specific enzymatic processes operating in the cells. Indicators of functional diversity are also indicators of microbial activity and thereby integrate diversity and function.

BIOLOG™ assay: Carbon utilisation patterns can be measured by the BIOLOG™ assay. In this assay, a soil extract is incubated with up to ninty five different carbon sources in a microtiter plate and a redox-dye is used to indicate microbial activity. Sets of specific carbon sources have been selected specifically for studies of soil microbial communities. The result of the assay is a qualitative physiological profile of the potential functions within the microbial community. Differences in the profiles can be analysed by multivariate statistics.

The BIOLOG™ assay is dependent on growth of cells under the specific conditions in the microtiter plate and thereby indicating only potential functional diversity. However, the technique has gained widespread use, primarily due to the ease of use and the capacity to produce comprehensive data sets. The data are analysed by multivariate statistics and experience with data interpretation is still developing. Considerable data are, however, available for a future reference database and this may facilitate data interpretation.

The BIOLOG™ assay has been shown to be more sensitive than microbial biomass and respiration measurements to impacts of soil management practices and of sewage sludge amendments to soil. The assay is currently implemented in the Dutch Soil Monitoring Programme where it has been shown to be discriminatory to different types of soil and management practices. The assay is also recommended for soil monitoring in Scotland and Northern Ireland. However, caution should be taken when using a commercial assay, as the product may go off the market or change composition. The utility of the assay can be extended by the Pollution-Induced Community Tolerance (PICT) approach, where a range of concentrations of a specific heavy metal can be added into the plate and the tolerance of the community be estimated.

Enzymatic activity: The enzymatic activity in soil is mainly of microbial origin, being derived from intracellular, cell-associated or free enzymes. Only enzymatic activity of ecto-enzymes and free enzymes is used for determination of the diversity of enzyme patterns in soil extracts. Discrimination between free and cell-associated enzyme activity can be obtained by a simple filtration step to separate microbial cells from the soil extract. The enzyme activity is quantified by incubation of the soil extract with commercial fluorogenic enzyme substrates (4-methylumbelliferin (MUF) and 4-methylcoumarinyl-7-amide (MC) or colometric substrates (remazol brilliant blue, p-nitrophenol or tetrazolium salt) coupled with specific compounds of interest (*e.g.* cellulose or phosphate). The data are typically analysed by multivariate statistics.

If incubation times are kept short, cell growth and synthesis of new enzymes are prevented. It has been recommended that a diverse set of enzyme activities are measured, since a few dominating organisms expressing a high enzyme activity may give a biased result. Similar diversity patterns were obtained when various soil types were tested by a set of 14 MUF-substrates and by the BIOLOG™ assay. Compared to the BIOLOG™ assay, this enzyme activity assay is closer to in situ functions, since it is independent of growth and enzyme synthesis. However, fewer functions are generally measured.

mRNA diversity: mRNA molecules are gene copies used for synthesis of specific proteins by the cell. The nucleotide sequences of mRNA molecules reflect the type of enzymes synthesised. Concentration of mRNA is correlated with the protein synthesis rate and as such with the activity of the microorganism. Therefore, the content and diversity of mRNA molecules will give very accurate pictures of the in situ function and activity of the microbial community.

Detection and quantification of a specific mRNA molecule can be done by reverse transcription PCR (RT-PCR), which is a very sensitive method. A prerequisite of this technique is knowledge of the nucleic acid sequence of the mRNA for a specific gene. For certain genes, this information is available. However, the technique of quantifying mRNA is still in its developmental stage. Sensitivity of the method has though been improved by associating a magnetic capture system.

Structural diversity

Phospholipid fatty acids (PLFAs) are stable components of the cell wall of most microorganisms. They are polar lipids specific for subgroups of microorganisms, *e.g.* gram-negative or gram-positive bacteria, methanotrophic bacteria, fungi, mycorrhiza, and actinomycetes. Individual PLFAs can thus be related to microbial community structure. The method gives a fingerprint of the relative PLFA composition of the resident microbial community.

PLFAs are extracted from soil samples and subsequently analysed by gas chromatography. Specific PLFAs are subsequently identified and/or quantified and the result is evaluated by multivariate statistics. PLFA profiles of soil samples offer sensitive reproducible measurements for characterising the numerically dominant portion of soil microbial communities without cultivating the organisms. The technique gives estimates of both microbial community composition and biomass size, and the results represent the in situ conditions in soil. The method is, however, timeconsuming, although the extraction procedure may be automated. PLFA analysis has been used to detect a pollution gradient in soil and found to be more discriminatory than BIOLOG™ measurements for characterising soil microbial communities. The method has recently been recommended for soil monitoring in Scotland and Northern Ireland.

Oligo- and copiotrophic bacteria

The ratio of oligotrophs to copiotrophs (bacteria that require a high nutrient input) has been proposed to reflect the nutrient stress tolerance of the species present in soil. A high ratio, *e.g.* dominance of oligotrophs, may indicate stable environmental conditions with low substrate availability. A low ratio, *e.g.* dominance of copiotrophs, may, in contrast, indicate an environment regularly receiving input of organic rich substrate, *e.g.* addition of sewage sludge or pesticides.

The ratio of oligotrophs to copiotrophs can be determined by either colony appearance on agar substrates, the rRNA-gene copy number in isolated bacteria or rRNA-expression in bacterial microcolonies. The appearance of colonies on agar substrates may simply be determined by counting colony forming units (CFUs) at specific time intervals. The counts are complemented by calculation of mean lag-phases and absolute numbers of bacterial subpopulations. Early appearing CFUs represent copiotrophic bacteria, while late appearing CFUs represent oligotrophic bacteria.

The number of rRNA copies in isolated bacteria, determined by molecular techniques, has recently been shown to correlate with the expression of the rRNA gene. The rRNA gene expression can be determined during growth in bacterial microcolonies (mCFUs) by measuring the 16S rRNA concentration by fluorogenic in situ hybridisation. A low rRNA-copy number or a low rRNA expression during growth indicate dominance of oligotrohic bacteria and hence a high ratio.

The CFU method is a simple and inexpensive method and is ready to use upon standardisation of incubation and counting procedures. The molecular methods are more comprehensive and time consuming and still need considerable testing before implementation into a monitoring programme. However, they have the potential for specifying the interacting groups of organisms depending on the specificity of the hybridisation probes in use.

Carbon Cycling

A major activity of soil microorganisms is decomposition of organic matter. Soil microorganisms are in general heterotrophic and rely on input of carbon energy from outside the microbial community. Organic matter in soil is largely derived from higher plants consisting of cellulose (15-60%), hemicellulose (10-30%) and lignin (5-

30%). Indicators of carbon cycling represents measurements at the ecosystem level.

Soil respiration

Soil respiration, which is the biological oxidation of organic matter to CO_2 by aerobic organisms, notably microorganisms, occupies a key position in the C cycle of all terrestrial ecosystems. It provides the principal means by which photosynthetically fixed carbon is returned to the atmosphere. The metabolic activities of soil microorganisms can be quantified by measuring CO_2 production and/or O_2 consumption. Measurement of soil respiration is one of the oldest, but still most frequently used techniques for quantification of microbial activities in soil. Soil respiration is positively correlated with soil organic matter content, and often with microbial biomass and microbial activity.

Soil respiration measurements are included in most soil monitoring programmes and have been found to discriminate between different soil types and land uses within the Dutch Soil Monitoring Programme. Soil respiration can be determined by either CO_2 production or O_2 consumption. Measurement of CO_2 concentration is more sensitive, because the atmospheric concentration of CO_2 is only 0.033% versus 20% for O_2.

Determination of CO_2 production from soil samples can be made in the laboratory by simple and inexpensive techniques based on alkaline CO_2 traps followed by chemical titration or by more sophisticated automated instruments based on electrical conductivity, gas chromatography or infrared spectroscopy. Combined with automated sampling from test soil samples, automated instruments make it possible to determine CO_2 production as a function of time for several days.

Respiration is highly influenced by temperature, soil moisture, nutrient availability and soil structure. Pre-conditioning and standardisation of the soil before measuring respiration is necessary to minimise the effect of these variables. Field measurements of soil respiration are less often used due to the high sensitivity to environmental conditions, although such measurements have been shown to discriminate between different soil management practices. Finally, soil respiration measurements have been used as an indicator of pesticide and heavy metal toxicity.

Metabolic quotient

The metabolic quotient (qCO_2), also called the specific respiratory rate, is defined as the microbial respiration rate (measured as evolution of CO_2) per unit microbial biomass. Microbial biomass for this purpose is often determined by substrate induced respiration, and the respiratory activity is determined concomitantly using the same instruments. The qCO_2 is included in the soil monitoring programme in Schleswig-Holstein, Germany. The qCO_2 has been used to study soil over time and, generally, the quotient decreases as the soil ages. Furthermore, the qCO_2 has been used in effect studies of environmental conditions, such as temperature and pH, soil management, soil texture and compaction and heavy metals.

Generally, the qCO_2 is found to be highest when ecosystem stress level is high. Caution, however, should be taken when interpreting qCO_2, since a high quotient may infer stress, an immature ecosystem or a more respirable substrate. Threshold values of qCO_2 for different soil texture classes of conventional agricultural soils have been elucidated within the soil monitoring programmes in Germany (Lower Saxony) and Switzerland.

Organic matter decomposition

Any disturbance in microbial activity will result in a change of the organic matter (OM) decomposition rate and hence the availability and cycling of the most important organic bound nutrients within the ecosystem, such as carbon, nitrogen, sulphur and phosphorus. Knowledge about rates of OM decomposition is thus a prerequisite for understanding the availability and recycling of all these nutrients. Field incubation of different types of plant litter or more standardised pieces such as cotton strips and wood sticks, are the most commonly used methods for studying OM decomposition rates.

Soil enzymes

Enzymes are the direct mediators for biological catabolism of soil organic and mineral components. Thus, these catalysts provide a meaningful assessment of reaction rates for important soil processes. Soil enzyme activities

- are often closely related to soil organic matter, soil physical properties and microbial activity or biomass,
- change much sooner than other parameters, thus providing early indications of changes in soil health, and
- involve simple procedures.

In addition, soil enzyme activities can be used as measures of microbial activity, soil productivity, and inhibiting effects of pollutants. Disturbance of the soil microbial activity, as shown by changes in levels of metabolic enzymes, can serve as an estimate of ecosystem disturbance. This relationship has been clearly shown when soil is polluted with heavy metals.

Easy, well-documented assays are available for a large number of soil enzyme activities. These include dehydrogenase, β-glucosidases, urease, amidases, phosphatases, arylsulphatase, cellulases and phenol oxidases. A standard method for determination of acid phosphatase activity exists within the ICP-IM soil monitoring network. Hydrolysis of the fluorescent fluorescein diacetate is thought to broadly represent soil enzyme activity, because it is hydrolysed by a number of different enzymes, such as proteases, lipases and esterases. These enzymatic activities are widely distributed in soil, where they mainly originate from microorganisms, but also from plants or animals.

Enzyme activities can be measured as in situ substrate transformation rates or as potential rates if the focus is more qualitative. An important parameter is whether decisions are made relative to in situ or to maximum enzyme activities. For comparisons of soil enzyme activities, the natural choice is the maximum activities. Measurements of soil enzyme reaction are usually based on the addition of an artificial, soluble substrate at a concentration sufficient to maintain zero-order kinetics, thus achieving a reaction rate proportional to enzyme concentration. Long incubation periods have to be omitted to avoid substrate depletion and microbial growth. Enzyme activities are usually determined by a dye reaction followed by a spectrophotometric measurement.

Methane oxidation

Methane (CH_4) is found extensively in Nature and is a

greenhouse gas in the atmosphere. Methane is produced in anoxic environments by methanogenic Archaea and consumed by aerobic methaneoxidising bacteria, the methanotrophs. Important terrestrial sites for methane oxidation are wetland areas receiving a high input of organic material. Furthermore, landfills containing high amounts of organic wastes are a source of methane and the habitat of many methanotrophs.

Net production of methane can be considered as an indicator of greenhouse gas emission and may further be linked to monitoring of the atmospheric balance. Methane oxidation is measured by spiking a soil sample with methane and incubate the sample in a closed jar in the laboratory. Loss of methane is subsequently determined by gas chromatography. The number of methanotrophs is an indicator of potential greenhouse gas consumption. Methanotrophs can be quantified directly in soil by fluorescent in situ hybridisation (FISH) or standard growth-dependent MPN counts. Analyses of methanotrophic communities can be done with PCR-DGGE using methanotrophs-specific 16S rDNA primers.

Nitrogen Cycling

The mineralisation of soil organic nitrogen (N) through nitrate to gaseous N_2 by soil microorganisms is a very important process in global N-cycling. This cycle includes N-mineralisation, nitrification, denitrification and N_2-fixation (Figure 1). Indicators of nitrogen cycling represent measurements at the ecosystem level. Organic N is mineralised to ammonium (NH_4^+) by a wide variety of soil microorganisms and it reflects the turnover of organic material in soil and the available indigenous N-pools to plants.

Ammonium is subsequently either immobilised by soil microorganisms (that is, assimilated into new biomass) or oxidised to nitrite (NO_2^-) and subsequently to nitrate (NO_3^-) by aerobic nitrification. Chemoautotrophic bacteria, the nitrifier population, carry out this process. At this step, leaching of N to the groundwater may occur due to the negative charge of the nitrate ion. Under normal circumstances, however, nitrate is subsequently reduced to gaseous nitrogen (N_2) via nitrous oxide (N_2O) by anaerobic denitrification. Denitrification is represented by a variety of soil bacteria.

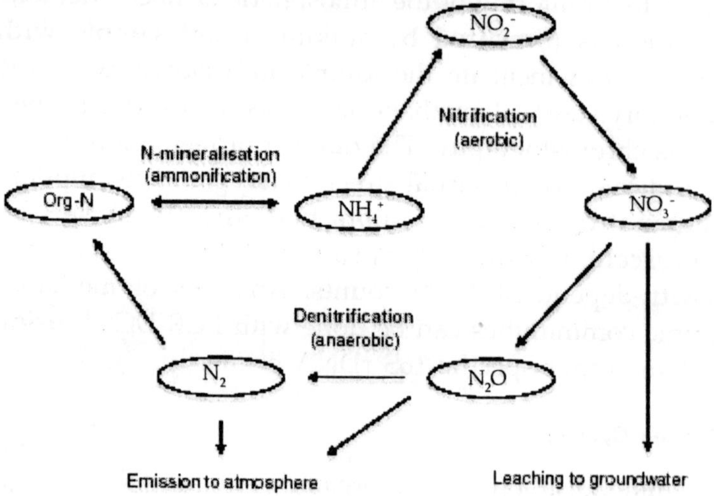

Figure 1. Nitrogen cycling

Nitrification and denitrification together lead to losses of bioavailable N since nitrous oxide and gaseous N_2 may be lost to the atmosphere. N_2 can be re-fixed into the soil by N_2-fixing microorganisms. Nitrous oxide is a greenhouse gas when lost to the atmosphere. subsequently either immobilised by soil microorganisms (that is, assimilated into new biomass) or oxidised to nitrite (NO_2^-) and subsequently to nitrate (NO_3^-) by aerobic nitrification. Chemoautotrophic bacteria, the nitrifier population, carry out this process. At this step, leaching of N to the groundwater may occur due to the negative charge of the nitrate ion. Under normal circumstances, however, nitrate is subsequently reduced to gaseous nitrogen (N_2) via nitrous oxide (N_2O) by anaerobic denitrification.

Denitrification is represented by a variety of soil bacteria. Nitrification and denitrification together lead to losses of bioavailable N since nitrous oxide and gaseous N_2 may be lost to the atmosphere. N_2 can be re-fixed into the soil by N_2-fixing microorganisms. Nitrous oxide is a greenhouse gas when lost to the atmosphere.

N-mineralisation

Ammonification is actually a measure of the net N-mineralisation, since immobilisation of NH_4 by soil microorganisms into new biomass occurs simultaneously with the mineralisation process. The measurement thus reflects the potential N-mineralisation in soil and is measured by the accumulation of NH_4 in soil slurry under aerobic conditions over a period of several weeks. Anaerobic incubation is sometimes preferred because there is less microbial immobilisation under anaerobic conditions and nitrification is inhibited.

Nitrification

Nitrification is believed to be a more sensitive parameter than Nmineralisation, because only a small number of bacteria, the nitrifiers, are involved in the process. Nitrification measurements are included in soil monitoring in Austria, the Czech Republic and an ISO-standard is available. Nitrification measurements have, however, been reported to be no more sensitive than N-mineralisation. Nitrification measurements reflect the population size of the nitrifiers since ammonium is an essential substrate for these organisms.

Furthermore, these measurements together with denitrification measurements may indicate deposition of ammonia on N-limited habitats. Nitrification is measured by the ammonium oxidising assay. With this method, a soil slurry is incubated with excess ammonium and chlorate, the latter inhibiting the oxidation of nitrite to nitrate. The oxidation of ammonium to nitrite is measured by gas chromatography.

Denitrification

The denitrifying capacity is a widespread feature among soil bacteria and therefore denitrification can be used as a representative for microbial biomass. Since denitrification is an anaerobic process the amount of denitrification found in soil is very dependent on abiotic factors such as precipitation and soil compaction. Thus, soil management practices readily influence the amount of denitrification found in agricultural fields. Denitrification measurements may, together with nitrification measurements, indicate deposition of ammonia in N-limited habitats. Measurement of denitrification is carried out by the acetylene inhibition technique, in which the reduction of N_2O to N_2 is inhibited by acetylene and accumulated nitrous oxide is measured by gas chromatography. Nitrate must be available in surplus.

The method is often used to measure the potential denitrification where nitrate and carbon are added and anaerobic conditions are established. However, interpretation of denitrification data is complicated, because the denitrification enzymes are synthesised only under anaerobic conditions and the enzymes are not functional under aerobic conditions, even though they persist in the microbial community. The denitrification assay may thus reflect historical anaerobic situations and not necessarily the size of the actively denitrifying biomass.

Soil Biomass

Biomass is fundamental for soil processes to occur and quantification of microbial biomass is as such a measurement at the ecosystem level.

Microbial biomass

Soil microbial biomass represents the fraction of the soil responsible for the energy and nutrient cycling and the regulation of organic matter transformation. A number of studies has reported a close relationship between soil microbial biomass, decomposition rate and N-mineralisation. Microbial biomass has also been shown to correlate positively with grain yield in organic, but not in conventional farming. Finally, soil microbial biomass contributes to soil structure and soil stabilisation. Soil microbial biomass has also been recommended as indicators of soil organic carbon.

Several methods have been used for the estimation of microbial biomass in soil. The methods can be divided into direct (*e.g.* microscopy or determinations of specific membrane phospholipid fatty acids (PLFAs)) and indirect (*e.g.* chloroform fumigation (CFE/CFI) or substrate induced respiration (SIR)). Microbial biomass measurements are used in several soil monitoring programmes: microscopy in the Dutch Soil Monitoring Programme, CFE or SIR in monitoring programmes in Germany, Switzerland, United Kingdom and The Czech Republic.

Determination of soil microbial biomass by direct methods (microscopy or PLFA analysis) gives results that very closely represent the in situ soil conditions. Although the methods are time-consuming, they are currently used for soil monitoring purposes. The automation of PLFA extraction has reduced analysis time to some extent. Direct counts or bio-volume estimations using conversion factors can estimate microbial biomass. Different soil preparation methods and staining techniques in combination with epifluorescens microscopy are available. A Danish standard for epifluorecens microscopy is further available.

Combined with automated image analysis, direct counts can be used routinely for the determination of soil microbial biomass in many samples of different origin. The total amount of PLFAs in soil is an alternative method to microscopic counting. PLFAs are found only in membranes of bacteria and fungi. Individual PLFAs are specific for specific subgroups of microorganisms. Using extraction of soil samples and analysis by gas chromatography, the total amount of PLFAs can be quantified. It is also possible to quantify different groups of microorganisms by this method. PLFA analysis hereby provides information on biodiversity and the fungalbacterial biomass ratio.

Indirect methods are generally cheaper, faster and easier to use than the direct methods. Results obtained by the indirect methods have been documented to be very close to the direct measurements, thus providing confidence in the utility of indirect methods. Chloroform fumigation is the most commonly used indirect method. This method is considered to measure most of the soil microbial biomass, *e.g.* both dead and alive; though some microorganisms (*e.g.* spores) are insensitive to the fumigation process.

Determination of microbial biomass by chloroform fumigation covers two indirect methods: the chloroform fumigation incubation method (CFI) and the chloroform fumigation extraction method (CFE). In both cases, the chloroform vapour kills the microorganisms in the soil, and subsequently the size of the killed biomass is estimated either by quantification of respired CO_2 over a specified period of incubation (CFI) or by a direct extraction of the soil immediately after the fumigation followed by a quantification of extractable carbon. The release of CO_2 after fumigation is the result of germinating microbial spores utilising the C substrate provided by the killed microbial cells.

Another common indirect method is substrate induced respiration (SIR). This method measures only the metabolically active portion of the microbial biomass. SIR measures the initial change in the soil respiration rate as a result of adding an easily decomposable substrate (*e.g.* glucose). The technique has been automated and is used in soil monitoring in several countries, *e.g.* Germany. Soil microbial biomass is subsequently calculated using a conversion factor.

The amount of microbial biomass carbon (C_{org}) may be related to the total carbon (C_{org}) content by the microbial quotient (C_{org}/C_{org}). This quotient provides a measure of soil organic matter dynamics and can be used as an indicator of net C loss or accumulation. Using the quotient avoids the problems of comparing trends in soils with different organic matter content. Living fungal biomass can be estimated by quantification of fungalspecific membrane molecules such as ergosterol or specific phospholipids (PLFAs).

The procedure for determination of ergosterol content in soil is simpler compared to determination of PLFAs. However, an important disadvantage of this method is that oomycetous fungi and a number of yeasts do not produce ergosterol. Additionally, it is recommended that total hyphal length is measured simultaneously for precise estimations of only living fungal biomass, but this is a very laborious and cumbersome technique. Quantification of enzyme activities such as fluorescein diacetate hydrolytic activity (FDA) or N-acetyl-beta-glucosaminidase (Nag) activity have been proposed as alternative, semi-quantitative measures of soil fungal biomass.

The fungal-bacterial biomass ratio can also be determined directly from measurements of fungal-specific and bacterial-specific PLFAs. More information is thus obtained from one single PLFA-analysis. The ratio has been used in soil management studies as a microbial indicator. A higher ratio is typical of long-term unfertilised or organic managed grasslands compared to fertilised grasslands of the same soil type.

Protozoan biomass

Protozoan biomass is determined by extracting a soil sample and counting directly by use of an inverted microscope. This yields the number of active protozoa. However, the vast majority of protozoa are encysted. An alternative method is thus to extract protozoa from the soil followed by a MPN counting based on a growth medium that causes protozoa to excyst. Both methods are very laborious and limited by the problems of extraction efficiency. The MPN approach further possesses the problems of culturability; not all cysts will excyst and not all protozoa grow under the laboratory conditions in liquid culture.

A newly developed molecular method, MPN-PCR, has been used to quantify a specific group of soil flagellates directly in a gnotobiotic soil system and higher but corresponding numbers was found compared to traditional MPN counting based on culturing. The application of MPN-PCR assays for soil protozoa is, however, currently limited by the scarcity of molecular data. Bioassays based on a 24 h growth response of common ciliates have been developed and successfully applied to heavy metal toxicity testing.

Microbial Activity Indicators

Indicators of microbial activity in soil represent measurements at the ecosystem level (*e.g.* processes regulating decomposition of organic residues and nutrient cycling, especially nitrogen, sulphur, and phosphorus). Measurements at the community level include bacterial DNA and protein synthesis. Frequency of bacteriophages is a measurement at the population level.

Synthesis of bacterial DNA

Synthesis of DNA is a prerequisite for bacterial cell division and, as such, an indicator of bacterial growth. DNA is unique in the way that it only participates in cell division. DNA synthesis can be determined by incorporation of ^{3}H- or ^{14}C-thymidine into bacterial DNA as thymidine is a unique nucleoside, which only participates in DNA synthesis. The method has several requirements:

(i) DNA synthesis has to be linearly correlated with the cell growth;

(ii) all bacteria must take up thymidine through the cell membrane, which has been shown not to be the case;

(iii) thymidine should not be metabolised; and

(iv) the radioactive label (^3H) should not exchange with other molecules, *e.g.* proteins. It has been shown that only 5-20% of the ^3H-thymidine incorporated into total macromolecules is incorporated into DNA.

A soil extract is incubated with radiolabelled thymidine for a short time period and then filtered to measure the amount of radiolabel in the cells. A thorough extraction and purification of DNA from the cells can solve the problem with unspecific incorporation of radiolabel. The method is extensively used in aquatic environments. During the last decade it has been adopted to soil, but the use is not as widespread as in aquatic environments. Bacterial growth rate (number of cells formed per unit time) is calculated by use of a conversion factor. This conversion factor is based on many assumptions, including estimates of the number of cells present and the amount of radiolabelled thymidine incorporated in relation to GC content of the total DNA content of cells.

Synthesis bacterial protein

Bacterial protein synthesis is directly correlated to bacterial activity and can be determined by incorporation of ^3H or ^{14}C leucine, as this amino acid is incorporated into proteins only. The method for leucine incorporation is the same as for thymidine incorporation and the incorporation of both precursors can be carried out in a single assay if different radiolabels are used. The advantages and drawbacks of the method are the same as for radiolabelled thymidine incorporation, although balanced growth is not a prerequisite. Furthermore, most bacteria take up leucine, although the incorporation efficiency may differ between soils. Measurements of protein synthesis are supposed to be more accurate than that of DNA synthesis, because of a relatively higher protein content in cells.

RNA measurements

The RNA molecules, ribosomal RNA (rRNA) and messenger RNA (mRNA), play key roles in the protein synthesis. The amount of RNA in individual cells or in a community may, therefore, be taken as an indicator of protein synthesis and, thus, microbial activity. The number of active cells can be detected by fluorescent in situ hybridisation (FISH). By this method, individual cells carrying high concentrations of rRNA, situated on ribosomes, are quantified by fluorescence microscopy. The amount of rRNA in a community can also be detected by Reverse Transcriptase Polymerase Chain Reaction (RT-PCR), where rRNA extracted from soil is detected by creating a DNA copy and separating by gel electrophoresis.

Quantification of activity by either method is still problematic and comprehensive method development is needed before implementation into a monitoring programme. In the future this will also include implementation of microarrays with simultaneous measurements of numerous genes. mRNA molecules are gene copies used for synthesis of specific proteins by the cell. Determination of mRNA can be taken as equivalent to the expression of a specific gene in soil. Such measurements can be done by real time quantitative RT-PCR, which detects and quantifies low amounts of mRNA in environmental samples including soil. A prerequisite for using this method is knowledge of the sequence of the mRNA. At present, this method is probably too advanced for use as a microbial indicator in a monitoring programme, but with further method development it may prove useful.

Bacteriophages

A bacteriophage is a virus, which infects and multiplies in a specific host bacterium. Bacteriophages are abundant in the soil environment and have been isolated for nearly every known species of soil bacteria. Most phages isolated from soil are temperate phages, *e.g.* phages that can lie dormant in bacterial cells after infection. The multiplication of bacteriophages strictly depends on the activity of the host bacteria. As such, monitoring of the frequency and host specificity of free bacteriophages in soil is an indicator of the activity of specific soil bacteria. This is in contrast to the other microbial activity indicators, which measure the activity of whole microbial communities.

Determination of free bacteriophages in soil can be carried out by a standard method of extraction followed by a plaque-assay with specific host bacteria, *e.g. Pseudomonas, Bacillus, Rhizobium*. A high number of plaques are presumed to indicate a recent high activity of similar host bacteria in the test soil assuming a direct correlation between the number of bacteriophages and bacterial activity. Such a correlation has indeed been shown for Azospirillum brasilense and Serratia liquefaciens, but has to be confirmed for other bacterial groups.

The selection of host bacteria should be representative for the soil type to be investigated. Furthermore, the bacteriophage sensitivity to the host bacteria should be known. The frequency and persistence of the bacteriophages in different soil types should be estimated a priori in order to standardise the method. Generally, temperate bacteriophages survive for long periods of time within the host bacteria. Without host bacteria, the survival of bacteriophages depends on abiotic parameters, *e.g.* clay content, soil moisture, temperature and pH.

Microbial Species in Soil

Microbial key species in soil are here defined as organisms that possess important functions in the soil ecosystem (*e.g.* nutrient cycling, plant pathogenesis) or are of human health concern (*e.g.* human pathogens/ zoonoses). A number of criteria has to be fulfilled for key species to be useful in a monitoring programme. For example, they should be (ecologically) relevant, preferably abundant, and easy to enumerate and identify. Key indicator species represent measurements at the population level.

Mycorrhiza

The majority of higher plants exist in natural symbiosis with mycorrhizal fungi. The group of mycorrhizal fungi includes ectomycorrhizal (mainly forest trees), arbuscular mycorrhizal (terrestrial plants) and ericoid mycorrhizal (heather) fungi. They colonise plant roots and provide the plant with nutrients, especially phosphorus, due to the increased nutrient availability caused by the extra-radical mycelium. Furthermore, mycorrhizal associations can have a positive influence on plant diversity, plant stress and disease tolerance, and on soil aggregation.

Colonisation by AM has been shown to be highly dependent on the presence of host plants, on land use and on soil management practices. Spore abundance and diversity have been shown to discriminate between extensively and intensively managed soils and AM diversity has been reported to be sensitive to heavy metal contamination, organic pollutants and atmospheric deposition. Quantitative analysis of AM based on spore morphology is implemented as a microbial indicator in the Swiss soil monitoring network, where it is used to indicate heavy metal contamination in soil. Colonisation of AM in soil has been proposed as an important indicator of plant and ecosystem health.

Abundance and diversity of AM is determined by extraction of spores from soil samples and subsequent counting in a microscope. An alternative method is to use the test soil in a plant bioassay and harvest either the spores or the roots. The determination of spore numbers is, however, poorly correlated to the actual colonisation potential of the soil and molecular tools for detection of AM in roots are the future needs within the Swiss Soil Monitoring Programme. Methods for direct detection and quantification of AM in soil samples or in roots have been developed. These include 18S rDNA PCR, nested PCR at the species level and AM-specific PLFAs.

Suppressive soil

Many of the proposed soil health indicators focus on the presence of beneficial rather than the absence of detrimental organisms, although both are important. The presence of plant pathogens (*e.g.* fungi) in soil may indicate the existence of other soil health problems, *e.g.* nutrient imbalance. A suppressive soil is able to suppress specific plant diseases by inherent biotic and abiotic factors. The suppressiveness of a certain soil may thus be an indicator of plant health. Several methods are available for determining soil suppressiveness. It can be determined by inoculation of target-plant seeds directly into the test soil or into a pathogen-infested test soil.

The frequency of diseased plants and/or pathogenic propagules in soil is scored after incubation for about three to four weeks and compared to a reference soil. The plant bioassay is a conventional technique and a positive correlation between the plant bioassay and the actual field measurements has been shown for suppressiveness of pea root ro. A specific test plant system has to be selected for a monitoring programme and the correlation between bioassay and field measurements has to be confirmed on a diverse set of soils. The assay requires a relatively long time before the results can be obtained, but it is simple and cheap.

Human pathogens

Human pathogens can enter agricultural soils through amendment with manure and sewage sludge. The presence of human pathogenic bacteria in soil is an indicator of potential human infection and as such an indicator of human health. Presence of Escherichia coli, have traditionally been used as an indicator of faecal contamination and hence as an indicator of the possible presence of other more pathogenic bacteria. Since the ability of the pathogenic bacteria to survive in the environment may not necessarily be equal to that of *E. coli*, it would be advantageous if the pathogens were enumerated directly.

Enumeration of pathogenic bacteria can be carried out either by cultivation or by molecular/immunological techniques. Methods relying on cultivation use growth media selective for specific groups of microorganisms, *i.e.* XLD agar for isolation of Salmonella and Shigella and MacConkey agar for isolation of coliforms. These methods are well-established, cheap, and easy to use. Molecular techniques may give a more accurate estimate of the population sizes, as they do not rely on growth of the bacteria. On the other hand they may detect dead bacteria as well as free DNA.

Among the molecular methods, that would be suitable for a monitoring programme, are quantitative PCR and specific fluorescent oligo-nucleotide probes. With immunological methods, specific antibodies are used instead of oligo-nucleotide probes and the detection limit can further be lowered when combined with immunomagnetic separation. The drawback of using both the molecular and immunological techniques is that they technically are more demanding than the traditional culturing methods. Little is known about the occurrence of pathogenic bacteria in agricultural soil and investigations on the differences between fields receiving manure and/or sewage sludge and untreated fields are needed prior to implementation into a monitoring programme.

Indicators of Bioavailability

Chemical compounds may often be adsorbed to soil particles, such as clay minerals, and made unavailable to the biota. The bioavailable concentration will be equal to or lower than the total chemically extractable concentration. From an environmental viewpoint, the bioavailable fraction of a chemical compound may be a more relevant parameter than the chemically extractable fraction. Microorganisms can measure the bioavailability of a chemical compound in soil. Indicators of bioavailability represent measurements at the community and population levels.

Biosensor bacteria

Biosensor bacteria are designed to respond to certain

stress situations (*e.g.* toxicity) through the use of reporter genes. Environmental relevant bacteria can be selected and genetically modified by fusing reporter genes (*e.g.* bioluminescence) to the genes of interest and thereby give a certain signal to a specific response. Ultimately, fibre optic linked membrane bound biosensor probes may facilitate in situ ecotoxicity monitoring of soil. Biosensor bacteria responding to mercury or chromate or zink are presently available. The zink biosensor bacteria have been used for soil monitoring purposes, where it was the most discriminative method. Commercial biosensor bacteria products for overall ecotoxicological analysis are available and Microtox.

Plasmid-containing bacteria

The frequency of plasmid-containing soil bacteria has been shown to be higher in polluted soils compared to agricultural soils, and to increase by addition of heavy metals to soil. Thus, measurement of numbers of plasmid-containing bacteria or numbers of plasmids in soil can be used as a general indicator of environmental contaminants. If numbers of plasmids increase at a site, an investigation to identify the stress factor can subsequently be initiated. Two different approaches can be used to assess the occurrence of plasmids in soil, the endogenous and the exogenous approach.

By the endogenous approach, plasmids are extracted from soil bacteria isolated on agar plates followed by a visualisation of the plasmids on agarose gels. By the exogenous approach, suitable plasmid free recipient bacteria are used as "fishing rods". The plasmid free bacteria are mixed with a soil sample and allowed time to pick up (by conjugation) naturally occurring plasmids from the indigenous bacteria. Plasmids are extracted and visualised as in the endogenous approach. A major disadvantage of the endogenous plasmid extraction procedure is that it only analyses the fraction of soil bacteria that grow on cultivation media. This step is eliminated in the exogenous plasmid isolation procedure. However, only conjugative and mobilisable plasmids may be isolated by this method. The frequency and variability in plasmid numbers in different soil types should be estimated in order to standardise the method.

Antibiotic resistant bacteria

Restricted use of antibiotics (*e.g.* growth promoters) in agriculture has reduced but not eliminated antibiotic resistant bacteria in livestock and food. Urban effluents, which also contain antibiotics, have been demonstrated to result in an increase in the number of antibiotic resistant bacteria in riverine environments. Antibiotic substances have been detected in outlets of sewage treatment plants, manure and agricultural fields. Although the measured concentrations of antibiotic substances are generally below the minimum inhibitory concentration (MIC) to microorganisms, they may nevertheless select for the outgrowth of resistant bacteria in the soil ecosystem.

Very little, however, is known about the occurrence of resistant microorganisms in agricultural soil. Heavy metal pollution may also indirectly select for antibiotic resistant bacteria, since a correlation between bacterial antibiotic resistance and mercury concentration in riverine sediments has been observed. Thus, monitoring antibiotic resistant bacteria in soil will not only allow an assessment of the potential risk of antibiotic resistant bacteria to humans (human health), but can also be used as an indicator of industrial and urban pollution. Enumeration of antibiotic resistant bacteria can be carried out either by cultivation and/or molecular techniques.

Methods relying on cultivation on selective growth media containing antibiotics are well established, cheap, and can easily be implemented in a monitoring programme. By use of these methods, not only can numbers of resistant bacteria be estimated, but the MIC and the breakpoint value may also be determined. This is necessary because an antibiotic concentration appropriate to distinguish between resistant and sensitive bacteria of one species, may not be applicable to another.

A well-known drawback of the cultivation methods is non-culturability of some bacteria. This can be overcome by molecular techniques, which estimate the population sizes of the resistance genes. PCR and molecular gene probe analysis can possibly be used to detect a specific resistance gene in a soil sample and to develop quantitative PCR methods. Since little is known about the occurrence of antibiotic resistant bacteria in agricultural soil, some baseline testing is required to investigate the possible differences between treated and untreated fields. Monitoring of antibiotic resistant bacteria may be complemented with measurements of bioavailable concentrations of antibiotics by use of biosensor bacteria or plasmid-containing bacteria.

Catabolic genes

When the degradation pathway of a chemical compounds (*e.g.* pesticides) is known, key enzymes and catabolic genes can be identified and quantified. The presence of degradable chemical compounds in a soil is presumed to provoke a higher incidence and expression of corresponding catabolic genes due to either growth of bacteria or the spreading of the catabolic genes to the microbial community. Catabolic genes may, however, also be present due to their involvement in the degradation of naturally occurring and related organic compounds.

The incidence of specific catabolic genes thus gives information on the ability of a soil to modify or degrade xenobiotic compounds. An elevated expression of the catabolic genes will, on the other hand, indicate a partial or complete degradation of the corresponding organic compound. Several methods have been proposed for determination of the incidence and expression of specific catabolic genes. These include conventional culturing of degradative microorganisms, activity measurements of specific degradative key enzymes, and molecular methods for detection of catabolic genes (*e.g.* PCR, qPCR) and measurements of their expression (*e.g.* mRNA, rRNA, biosensor bacteria).

The molecular methods are described elsewhere and only the culturing technique will be dealt with here. The potential for degradation of a xenobiotic compound in soil can be estimated by incubation of a soil slurry spiked with the compound (radiolabelled or unlabelled) of interest and subsequent determinations of either radiolabelled CO_2-production, the respiration rate or cell growth. The incubation approach is also used for isolation of consortia or pure cultures able to grow on and degrade specific xenobiotic compound. The assay, though, is entirely dependent on the activity of the microorganisms and their culturability at the incubation conditions provided.

7

Soil Conservation

Soil conservation is a set of management strategies for prevention of soil being eroded from the earth's surface or becoming chemically altered by overuse, salinisation, acidification, or other chemical soil contamination. The principal approaches these strategies take are:

- choice of vegetative cover,
- erosion prevention,
- salinity management,
- acidity control,
- encouraging health of beneficial soil organisms,
- prevention and remediation of soil contamination,
- mineralisation.

Many scientific disciplines are involved in these pursuits, including agronomy, hydrology, soil science, meteorology, microbiology, and environmental chemistry.

Cover Crop

Decisions regarding appropriate crop rotation, cover crops, and planted windbreaks are central to the ability of surface soils to retain their integrity, both with respect to erosive forces and chemical change from nutrient depletion. Crop rotation is simply the conventional alternation of crops on a given field, so that nutrient depletion is avoided from repetitive chemical uptake/ deposition of single crop growth. Cover crops serve the function of protecting the soil from erosion, weed establishment or excess evapotranspiration; however, they may also serve vital soil chemistry functions.

For example, legumes can be ploughed under to augment soil nitrates, and other plants have the ability to metabolise soil contaminants or alter adverse pH. The cover crop Mucuna pruriens (velvet bean) has been used in Nigeria to increase phosphorus availability after application of rock phosphate. Some of these same precepts are applicable to urban landscaping, especially with respect to ground-cover selection for erosion control and weed suppression.

Soil pH

Soil pH levels adverse to crop growth can occur naturally in some regions; it can also be induced by acid rain or soil contamination from acids or bases. The role of soil pH is to control nutrient availability to vegetation. The principal macronutrients (calcium, phosphorus, nitrogen, potassium, magnesium and sulphur) prefer neutral to slightly alkaline soils. Calcium, magnesium and potassium are usually made available to plants via cation exchange surfaces of organic material and clay soil surface particles.

While acidification increases the initial availability of these cations, the residual soil moisture concentrations of nutrient cations can fall to alarmingly low levels after initial nutrient uptake. Moreover, there is no simple relationship of pH to nutrient availability because of the complex combination of soil types, soil moisture regimes and meteorological factors. The important observation is that pH is the regulatory mechanism to plant nutrient uptake, and that the theoretical concentration of soil nutrients is meaningless until pH levels are in the optimum range for uptake.

Soil pH can be raised by amendment by agricultural lime; The pH of an alkaline soil is lowered by adding sulfur, iron sulphate or aluminum sulphate, although these tend to provide costly short term benefits. Urea, urea phosphate, ammonium nitrate, ammonium phosphates, ammonium sulphate and monopotassium phosphate also reduce soil pH.

Soil Organisms

Promoting the viability of beneficial soil organisms is an element of soil conservation; moreover this includes macroscopic species, notably the earthworm, as well as microorganisms. Positive effects of the earthworm are known well, as to aeration and promotion of macronutrient availability. When worms excrete egesta in the form of casts, a balanced selection of minerals and plant nutrients is made into a form accessible for root uptake. US research shows that earthworm casts are five times richer in available nitrogen, seven times richer in available phosphates and eleven times richer in available potash than the surrounding upper 150 mm of soil. The weight of casts produced may be greater than 4.5 kg per worm per year. By burrowing, the earthworm is of value in creating soil porosity, creating channels enhancing the processes of aeration and drainage.

Soil microorganisms play a vital role in macronutrient cycles. For example, nitrogen fixation is carried out by free-living or symbiotic bacteria. These bacteria have the nitrogenase enzyme that combines gaseous nitrogen with hydrogen to produce ammonia, which is then further converted by the bacteria to make other organic compounds. Some nitrogen fixing bacteria such as rhizobia live in the root nodules of legumes. Here they form a mutualistic relationship with the plant, producing ammonia in exchange for carbohydrates.

In the case of the carbon cycle, carbon is transferred within the biosphere as heterotrophs feed on other organisms. This process includes the uptake of dead organic material (detritus) by fungi and bacteria in the form of fermentation or decay phenomena. Some soil microorganisms known as extremophiles have remarkable properties of adaptation to extreme environmental conditions including temperature, pH and water deprivation. The viability of soil organisms can be compromised when insecticidess and herbicides are applied to planting regimes. Often there are unforeseen and unintended consequences of such chemical use in the form of death of impaired functioning of soil organisms. Thus any use of pesticides or herbicides should only be undertaken after thorough understanding of residual toxicities upon soil organisms as well as terrestrial ecological components.

Killing soil microorganisms is a deleterious impact of slash and burn agricultural methods. With the surface temperatures generated, virtual annilation of soil and vegetative cover organisms are destroyed, and in many environments these effects can be virtually irreversible (at least for generations of mankind. Shifting cultivation is also a farming system that often employs slash and burn as one of its elements.

Mineralisation

To allow plants full realisation of their phytonutrient

potential, active mineralisation of the soil is sometimes undertaken. This can be in the natural form of adding crushed rock or can take the form of chemical soil supplement. In either case the purpose is to combat mineral depletion of the soil. There are a broad range of minerals that can be added including common substances such as phosphorus and more exotic substances such as zinc and selenium. There is extensive research on the phase transitions of minerals in soil with aqueous contact.

The process of flooding can bring significant bedload sediment to an alluvial plain. While this effect may not be desirable if floods endanger life or if the eroded sediment originates from productive land, this process of addition to a floodplain is a natural process that can rejuvenate soil chemistry through mineralisation and macronutrient addition.

Soil Contamination

Excavation of leaking underground storage tank causing soil contamination. Soil contamination is the presence of man-made chemicals or other alteration to the natural soil environment. This type of contamination typically arises from rupture of underground storage tanks, application of pesticides and herbicides, percolation of contaminated surface water to subsurface strata, leaching of wastes from landfills or direct discharge of industrial wastes to the soil. The most common chemicals involved are petroleum hydrocarbons, solvents, pesticides, lead and other heavy metals. The occurrence of this phenomenon is correlated with the degree of industrialisation and intensity of chemical usage.

The United States, while having some of the most widespread soil contamination, has actually been a leader in defining and implementing standards for cleanup.

Other industrialised countries have a large number of contaminated sites, but lag behind the U.S. in executing remediation. Developing countries may be leading today in the next generation of new soil contamination cases. The concern over soil contamination stems primarily from health risks, both of direct contact and from secondary contamination of water supplies.

Mapping of contaminated soil sites and the resulting cleanup are time-consuming and expensive tasks, requiring extensive efforts in the areas of geology, hydrology, chemistry and computer modeling. Each year in the U.S., thousands of sites complete soil contamination cleanup, many by using microbes that "eat up" toxic chemicals in soil, many others by simple excavation and others by more expensive high-tech soil vapor extraction or stripper tower technology. At the same time, efforts proceed worldwide in creating and identifying new sites of soil contamination, particularly in industrial countries other than the U.S., and in developing countries which lack the money and the technology to adequately protect soil resources.

To understand the fundamental nature of soil contamination, it is necessary to envision the variety of mechanisms for pollutants to become entrained in soil. Soil particulates may be composed of a gamut of organic and inorganic chemicals with variations in cation exchange capacity, buffering capacity, and poise capacity. For example, at the extremes, one has a sand component, a coarse grained, inert, and totally inorganic substance; whereas peat soils are dominated by a fine organic material, made of decomposing organic material and highly active.

Most soils are mixtures of soil subtypes and thus have quite complex characteristics. There is also a great diversity of soil porosity, ranging from gravels to sands to silt to clay (in increasing order of porosity), pore size, and pore tortuosity (both in decreasing order). Finally there is a wide spectrum of chemical bonding or adhesion characteristics: each contaminant has a different interaction or bonding mechanism with a given soil type.

On balance, some contaminants may literally drain through soils such as sand and gravel and move to other soils or deeper aquifers, while polar or organic chemicals discharged into a clay soil will have a very high adsorption. Thus most soil contamination is the result of pollutants adhering to the soil particle surface, or lodging in interstices of a soil matrix. Clearly the equilibrium reached is a dynamic one, where new pollutants may lodge on new soil particles and the action of groundwater movement may over time transport some of the soil contaminants to other locations or depths.

There are legions of contaminant sources and pollutant types, but the following list is illustrative (pollutants indicated by parentheses):

- Petroleum hydrocarbons from rupture of underground storage tanks (benzene, toluene, xylene, alkanes, alkenes and MTBE).
- Spillage or leakage of solvents and dry cleaning agents (acetone, trichloroethylene, formaldehyde) and perchloroethylene.
- Leaching of contaminants from solid waste disposal sites (lead, mercury, chromium, cadmium, bacteria, and hydrocarbons).
- Water runoff which carries pollutants and may deposit them at a point of percolation.
- Percolation into soils from pesticides and herbicides uses (wide variety of chemicals including DDT,

lindane, organochlorines, organophosphates, carbamates and cyclodienes).

- Deposition of dust from smelting operations and coal burning power plants (zinc, cadmium, lead and mercury).
- Lead deposition from lead abatement or construction demolition (lead).
- Leakage of transformers [Polychlorinated Biphenyls (PCBs)].

The major concern is that there are many sensitive land uses where people are in direct contact with soils such as residences, parks, schools and playgrounds. Other contact mechanisms include contamination of drinking water or inhalation of soil contaminants which have vaporised. There is a very large set of health consequences from exposure to soil contamination depending on pollutant type, pathway of attack and vulnerability of the exposed population.

Chromium and many of the pesticide and herbicide formulations are carcinogenic to all populations. Lead is especially hazardous to young children, in which group there is a high risk of developmental damage to the brain and nervous system, while to all populations kidney damage is a risk. Chronic exposure to benzene at sufficient concentrations is known to be associated with higher incidence of leukemia. Mercury and cyclodienes are known to induce higher incidences of kidney damage, some irreversible. PCBs and cyclodienes are linked to liver toxicity.

Organophosphates and carbamates can induce a chain of responses leading to neuromuscular blockage. Many chlorinated solvents induce liver changes, kidney changes and depression of the central nervous system. There is an entire spectrum of further health effects such as headache, nausea, fatigue, eye irritation and skin rash for the above cited and other chemicals. Clearly at sufficient dosages a large number of soil contaminants cause death.

Effects on ecosystems

Not unexpectedly, soil contaminants can have significant deleterious consequences for ecosystems. There are radical soil chemistry changes which can arise from the presence of many hazardous chemicals even at low concentration of the contaminant species. These changes can manifest in the alteration of metabolism of endemic microorganisms and arthropods resident in a given soil environment. The result can be virtual eradication of some of the primary food chain, which in turn have major consequences for predator or consumer species. Even if the chemical effect on lower life forms is small, the lower pyramid levels of the food chain may ingest alien chemicals, which normally become more concentrated for each consuming rung of the food chain.

Many of these effects are now well known, such as the concentration of persistent DDT materials for avian consumers, leading to weakening of egg shells, increased chick mortality and potentially species extinction. Effects occur to agricultural lands which have certain types of soil contamination. Contaminants typically alter plant metabolism, most commonly to reduce crop yields. This has a secondary effect upon soil conservation, since the languishing crops cannot shield the earth's soil mantle from erosion phenomena. Some of these chemical contaminants have long half-lives and in other cases derivative chemicals are formed from decay of primary soil contaminants.

Until about 1970 there was little widespread awareness of the worldwide scope of soil contamination or its health risks. In fact, areas of concern such as Love Canal were often viewed as unusual or isolated incidents. In the U.S., passage of the National Environmental Policy Act in 1969 required careful analysis of the consequences of any federally funded project. Passage of The Resource Conservation and Recovery Act (RCRA) by the U.S. Congress in 1976 established guidelines not only for handling of hazardous materials but transport and hauling, such as required in cleanup of soil contaminants.

In 1980 the U.S. Comprehensive Emergency Response Compensation and Liability Act (CERCLA) was passed to establish, for the first time, strict rules on legal liability for soil contamination. Not only did CERCLA stimulate identification and cleanup of thousands of sites, but it raised awareness of property buyers and sellers to make soil contamination a focal issue of land use and management practices. While estimates of remaining soil cleanup in the U.S. may exceed 200,000 sites, in other industrialised countries there is a lag of identification and cleanup functions. Lesser developed countries are not without a share of the creation of soil contamination.

Even though their use of chemicals is far less than industrialised countries, often their controls and regulatory framework is quite weak. For example, some persistent pesticides banned in the U.S. for decades are in widespread uncontrolled use in developing countries. It is worth noting that the cost of cleaning up a soil contaminated site can range from as little as about $10,000 for a small spill, which can be simply excavated, to millions of dollars for a widespread event, especially for a chemical that is very mobile such as MTBE or perchloroethylene.

Remediation for soil contamination

Cleanup or remediation is analysed by environmental scientists who utilise field measurement of soil chemicals and also apply computer models for analysing transport and fate of soil chemicals. Thousands of soil contamination cases are currently in active cleanup across the U.S. as of 2006. There are several principal strategies for remediation:

- Excavate soil and remove it to a disposal site away from ready pathways for human or sensitive ecosystem contact. This technique also applies to dredging of bay muds containing toxins.
- Aeration of soils at the contaminated site (with attendant risk of creating air pollution)
- Bioremediation, involving microbial digestion of certain organic chemicals. Techniques used in bioremediation include landfarming, biostimulation and bioaugmentation soil biota with commercially available microflora.
- Extraction of groundwater or soil vapor with an active electromechanical system, with subsequent stripping of the contaminants from the extract.
- Containment of the soil contaminants (such as by capping or paving over in place).

Retrogression and Degradation

Retrogression and degradation are two regressive evolution processes associated with the loss of equilibrium of a stable soil. Retrogression is primarily due to erosion and corresponds to a phenomenon where succession reverts back to pioneer conditions (such as bare ground). Degradation is an evolution, different of natural evolution, related to the locale climate and vegetation. It is due to the replacement of the primitive vegetation (known as climax) by a secondary vegetation. This replacement modifies the humus composition and amount, and impacts the formation of the soil. It is directly related to human activity.

Land degradation is a human induced or natural process which negatively affects the capacity of land to function effectively within an ecosystem by accepting, storing and recycling water, energy, and nutrients. Desertification is land degradation occurring in arid, semiarid and dry subhumid areas of the world. These susceptible drylands cover 40 percent of the earth's surface and puts at risk more than 1 billion people who are dependent on these lands for survival. Land degradation causes losses to agricultural productivity in many parts of the world. The causes of land degradation are mainly anthropogenic and mainly agriculture related. The major causes include:

– Land clearing and deforestation,
– Agricultural depletion soil nutrients,
– Urban conversion,
– Irrigation, and
– Pollution.

The major stresses on vulnerable land include:

– Accelerated soil erosion by wind and water,
– Soil acidification or alkalinisation,
– Salination, and
– Destruction of soil structure including loss of organic matter.

Severe land degradation affects a significant portion of the earth's arable lands, decreasing the wealth and economic development of nations. Land degradation cancels out gains advanced by improved crop yields and reduced population growth. As the land resource base becomes less productive, food security is compromised and competition for dwindling resources increases, the seeds of potential conflict are sown. Thus a downward eco-social spiral is created when marginal lands are nutrient depleted by unsustainable land management practices resulting in lost soil stability leading to permanent damage.

We often assume that land degradation only affects soil productivity. However, the effects of land degradation often have more significant impacts on receiving water courses (rivers, wetlands and lakes) since soil, along with nutrients and contaminants associated with soil, are delivered in large quantities to environments that respond detrimentally to their input. Land degradation therefore has potentially disastrous impacts on lakes and reservoirs that are designed to alleviate flooding, provide irrigation, and generate Hydro-Power.

Impacts of Ecological Factors on Soil Formation

There are two types of ecological factors influencing the evolution of a soil (through alteration and humification). These two factors are extremely significant to explain the evolution of soils of short development.

- A first type of factor is the average climate of an area and the vegetation which is associated (biome). This factor allows one to define the world major areas of vegetation and soils.

- A second type of factor is more local, and is related to the original rock and local drainage. This type of factor explains appearance of specialised associations (ex peat bogs).

Biorhexistasy Theory

The destruction of the vegetation implies the destruction of evoluted soils, or a regressive evolution. Cycles of succession-regression of soils follow one another within short intervals of time (human actions) or long intervals of time (climate variations). The climate role in the deterioration of the rocks and the formation of soils lead to the formulation of the theory of the biorhexistasy.

- In wet climate, the conditions are favorable to the deterioration of the rocks (mostly chemically), the development of the vegetation and the formation of soils; this period favorable to life is called biostasy.
- In dry climate, the rocks exposed are mostly subjected to mechanical disintegration which produces coarse detrital materials: this is referred to as rhexistasy.

Perturbations of the Balance of a Soil

When the state of balance, characterised by the ecosystem climax is reached, it tends to be maintained stable in the course of time. The vegetation installed on the ground provides the humus and ensures the ascending circulation of the matters. It protects the ground from erosion by playing the role of barrier (for example, protection from water and wind). Plants can also reduce erosion by binding the particles of the ground to their roots. A disturbance of climax will cause retrogression, but, if given the opportunity, nature will make every effort to restore the damage via secondary succession.

Secondary succession is much faster than primary because the soil is already formed, although deteriorated and needing restoration as well. However, when a significant destruction of the vegetation takes place (of natural origin such as an avalanche or human origin), the disturbance undergone by the ecosystem is too important. In this latter case, erosion is responsible for the destruction of the upper horizons of the ground, and is at the origin of a phenomenon of reversion to pioneer conditions.

The phenomenon is called retrogression and can be partial or total (in this case, nothing remains beside bare rock). For example, the clearing of an inclined ground, subjected to violent rains, can lead to the complete destruction of the soil. Man can deeply modify the evolution of the soils by direct and brutal action, such as clearing, abusive cuts, forest pasture, litters raking. The climax vegetation is gradually replaced and the soil modified (example: replacement of leafy tree forests by moors or pines plantations). Retrogression is often related to very old human practices.

Impacts of Human

Erosion is the main factor for soil degradation and is due to several mechanisms: water erosion, wind erosion, chemical degradation and physical degradation. Erosion is strongly related to human activity. For example, roads which increase impermeable surfaces lead to streaming and ground loss. Agriculture also accelerates soil erosion (increase of field size, correlated to hedges and ditches removal). Meadows are in regression to the profit of plowed lands. Spring cultures (sunflower, corn, beet) surfaces are increasing and leave the ground naked in winter. Sloping grounds are gradually colonised by vine.

Lastly, use of herbicides leaves the ground naked between each crop. New cultural practices, such as mechanisation also increases the risks of erosion. Fertilisation by mineral manures rather than organic manure gradually destructure the soil. Many scientists observed a gradual decrease of soil organic matter content in soils, as well as a decrease of soil biological activity (in particular, in relation to chemical uses). Lastly, deforestation, in particular, is responsible for degradation of forest soils. Agriculture increases the risk of erosion through its disturbance of vegetation by way of:

- overgrazing of animals,
- planting of a monoculture,
- row cropping,
- tilling or plowing,
- crop removal,
- land-use conversion,

Consequences of Land Degradation

- *Yields impact*: Recent increases in the human population have placed a great strain on the world's soil systems. More that 5.5 billion people are now using about 10 % of the land area of the Earth to raise crops and livestock. Many soils suffer from various type of degradation, that can ultimately reduce their ability to produce food resources. Slight degradation refers to land where yield potential has been reduced by 10%, moderate degradation refers to a yield decrease from 10-50 %. Severely degraded soils have lost more than 50% of their potential. Most severely degraded soils are located in developing countries such as Asia and Africa.

- *Natural disasters*: mud flows, floods... responsible for the death of many living beings each year.
- *Deterioration of the water quality*: the increase in the turbidity of water and the contribution of nitrogen and of phosphorus can result in eutrophication. Soils particles in surface waters are also accompanied by agricultural inputs and by some pollutants of industrial, urban and road origin (such as heavy metals). The ecological impact of agricultural inputs (such as weed killer) is known but difficult to evaluate because of the multiplicity of the products and their broad spectrum of action.
- *Biological diversity*: soil degradation may involve the disappearance of the climax vegetation, the decrease in animal habitat, thus leading to a biodiversity loss and animal extinction...

Problems of soil erosion can be fought, and certain practices can lead to soil enhancement and rebuilding. Even though simple, methods for reducing erosion are often not chosen because these practices outweigh the short-term benefits. Rebuilding is especially possible through the improvement of soil structure, addition of organic matter and limitation of runoff. However, these techniques will never totally succeed to restore a soil (and the fauna and flora associated to it) that took more than one thousand years to build up.

Soil Salination

Soil salination is the accumulation of free salts to such an extent that it leads to degradation of soils and vegetation. Salination is a natural process that results from:

- high levels of salt in the soils,

- landscape features that allow salts to become mobile (movement of water table),
- climatic trends that favor accumulation.

Salt is a natural element of soils and water. The ions responsible for salination are: Na^+, K^+, Ca^{2+}, Mg^{2+} and Cl^-. In some areas (for example in Australia), salinity is an inherent situation (enormous amounts of salts are stored in the soils). However, human practices have increased soil salinity by changing the natural balance of the water cycle, by allowing excess recharging of groundwater and accumuiation through concentration. One of the best examples of excess salination was observed in Egypt in 1970 when the Aswan High Dam was built.

The change in the level of ground water before the construction had enabled soil erosion, which led to high concentration of salts in the water table. After the construction, the continuous high level of the water table led to the salination of the arable land. Salinity from drylands can occur when the water table is between two to three metres from the surface of the soil. The salts from the groundwater are raised by capillary action to the surface of the soil. This occurs when groundwater is saline (which is true in many areas), and is favored by land use practices allowing more rainwater to enter the aquifer than it could accommodate.

For example, the clearing of trees for agriculture is a major reason for drylands in some areas, since deep rooting of trees has been replaced by shallow rooting of annual crops. Salinity from irrigation can occur over time wherever irrigation occurs, since almost all water (other than natural rainfall) contains some dissolved salts. When the plants use the water, the salts are left behind in the soil and eventually begin to accumulate. Since soil salinity makes it more difficult for plants to absorb soil moisture, these salts must be leached out of the plant root zone by applying additional water. This, in turn can lead to rising water tables, requiring drainage to keep the saline groundwater out of the root zone.

If the water table rises too high, then natural soil evaporation will begin to draw the salts back upward into the soil profile. The problem is accelerated when too much water is added too quickly due to inefficient water use such as over-irrigation, applying more than is required for leaching, using bad estimates of evapotranspiration and poor system design, and is also greatly increased by poor drainage and use of saline water for irrigating agricultural crops. These practices result in the concentration of salts. Salinity in urban areas often results from the combination of irrigation and groundwater processes. Cities are often located on drylands, leaving the rich soils for agriculture. Irrigation is also now common in cities (gardens and recreation areas).

The consequences of salinity are:

- detrimental effects on plant growth and final yield,
- damage to infrastructure (roads, bricks, corrosion of pipes and cables),
- reduction of water quality for users, sedimentation problems, and
- soil erosion ultimately, when crops are too strongly affected by the amounts of salts.

Salinity is an important land degradation problem. One way to prevent excess salination would be the use of humic acids, especially in regions where too much irrigation was practiced. In soils with excess salts, humic acids can fix anions and cations and eliminate them from the root regions of the plants. A type of plant that is of particular interest for those in areas affected by soil salinity is the saltbush, which is able to tolerate saline conditions and draws salt up into its leaves.

Desertification

Desertification is the degradation of land in arid, semi arid and dry sub-humid areas resulting from various factors including climatic variations and human activities. Modern desertification often arises from the demands of increased populations that settle on the land in order to grow crops and graze animals. However, there has been major controversy over the definition of the term, as well as its causes. Attempts to define or map areas that are or have turned into desert have been sharply criticised.

What is usually less controversial is the fact of biodiversity loss, and loss of productive capacity, such as the transition from grassland dominated by perennial grasses to one dominated by perennial shrubs. In the southwestern deserts of the United States, semiarid ecosystems dominated by perennial bunchgrasses, including blue grama and black grama, have been replaced by shrublands dominated by creosotebush since the early 1900s. The change in vegetation is thought to have induced desertification in this region.

In some regions around the world, deserts are separated sharply from surrounding, less arid areas by mountains and other contrasting landforms that reflect basic structural differences in the regional geology. In other areas, desert fringes form a gradual transition from a dry to a more humid environment, making it more difficult to define the desert border. These transition zones have very fragile, delicately balanced ecosystems. Desert fringes often are a mosaic of microclimates.

Small hollows support vegetation that picks up heat from the hot winds and protects the land from the prevailing winds. After rainfall the vegetated areas are distinctly cooler than the surroundings. In these marginal areas human activity may stress the ecosystem beyond its tolerance limit, resulting in degradation of the land. By pounding the soil with their hooves, livestock compact the substrate, increase the proportion of fine material, and reduce the percolation rate of the soil, thus encouraging erosion by wind and water.

Grazing and the collection of firewood reduce or eliminate plants that help to bind the soil. In large desert areas, sand dunes can encroach on human habitats. Sand dunes move through a few different means, all of them helped along by wind. One way that dunes can move is through saltation, where sand particles skip along the ground like a rock thrown across a pond might skip across the water's surface. When these skipping particles land, they may knock into other particles and cause them to skip as well.

With slightly stronger winds, particles collide in mid-air, causing sheet flows. In a major dust storm, dunes may move tens of meters through such sheet flows. And like snow, sand avalanches, falling down the steep slopes of the dunes that face away from the winds, also moving the dunes forward. It is a common misconception that droughts cause desertification. Droughts are common in arid and semiarid lands. Well-managed lands can recover from drought when the rains return.

Continued land abuse during droughts, however, increases land degradation. Increased population and livestock pressure on marginal lands has accelerated desertification. In some areas, nomads moving to less arid areas disrupt the local ecosystem and increase the rate of erosion of the land. Nomads are trying to escape the desert, but because of their land-use practices, they are bringing the desert with them. Some arid and semi-arid lands can support crops, but additional pressure from greater populations or decreases in rainfall can lead to the few plants present disappearing.

The soil becomes exposed to wind, causing soil particles to be deposited elsewhere. The top layer becomes eroded. With the removal of shade, rates of evaporation increase and salts become drawn up to the surface. This is salinisation, and inhibits plant growth. The loss of plants causes less moisture to be retained in the area, which may change the climate pattern leading to lower rainfall. This degradation of formerly productive land is a complex process. It involves multiple causes, and it proceeds at varying rates in different climates. Desertification may intensify a general climatic trend toward greater aridity, or it may initiate a change in local climate.

Desertification does not occur in linear, easily mappable patterns. Deserts advance erratically, forming patches on their borders. Areas far from natural deserts can degrade quickly to barren soil, rock, or sand through poor land management. The presence of a nearby desert has no dlrect relationship to desertification. Unfortunately, an area undergoing desertification is brought to public attention only after the process is well underway. Often little or no data are available to indicate the previous state of the ecosystem or the rate of degradation. Scientists still question whether desertification, as a process of global change, is permanent or how and when it can be halted or reversed.

Desertification is a historic phenomenon; the world's great deserts were formed by natural processes interacting over long intervals of time. During most of these times, deserts have grown and shrunk independent of human activities. Paleodeserts, large sand seas now inactive because they are stabilised by vegetation, extend well beyond the present margins of core deserts, such as the Sahara. Many deserts have come about around western asia because of an overpopulation of prehistoric species and subspecies during the late Cretaceous era. Through dated fossil pollen, it has been found that today's Sahara desert has been changing between desert and fertile savanna. Studies also show that advance and retreat depend on yearly rainfall.

A number of solutions have been tried in order to reduce the rate of desertification and regain lost land. Leguminous plants, which extract nitrogen from the air and fix it in the soil, can be planted to restore fertility. Stones stacked around the base of trees collect morning dew and help retain soil moisture. Artificial grooves can be dug in the ground to retain rainfall and trap wind-blown seeds. In Iran, petroleum is being sprayed over semi-arid land with crops. This coats seedlings to prevent moisture loss and stop them being blown away.

Windbreaks made from trees and bushes to reduce soil erosion and evapotranspiration was widely encouraged by development agencies from the middle of the 1980s in the Sahel area of Africa. With many of the local people using trees for firewood and cooking the problem has become acute. In order to gain further supplies of fuel the local population add more pressure to the depleted forests; thus adding to the desertification process. Solar ovens are being advocated as a means to relieving some of this pressure upon the environment.

While desertification has received tremendous publicity by the political and news media, there are still many things that we don't know about the degradation of productive lands and the expansion of deserts. At the local level, individuals and governments can help to reclaim and protect their lands. In areas of sand dunes, covering the dunes with large boulders or petroleum will interrupt the wind regime near the face of the dunes and prevent the sand from moving. Sand fences are used throughout the Middle East and the United States, in the same way snow fences are used in the north.

Placement of straw grids, each up to a square meter in area, will also decrease the surface wind velocity. Shrubs and trees planted within the grids are protected by the straw until they take root. However, some studies suggest that planting of trees actually depletes water supplies in the area. In areas where some water is available for irrigation, shrubs planted on the lower one-third of a dune's windward side will stabilise the dune. This vegetation decreases the wind velocity near the base of the dune and prevents much of the sand from moving.

Higher velocity winds at the top of the dune level it off and trees can be planted atop these flattened surfaces. Oases and farmlands in windy regions can be protected by planting tree fences or grass belts. Sand that manages to pass through the grass belts can be caught in strips of trees planted as wind breaks 50 to 100 meters apart adjacent to the belts. Small plots of trees may also be scattered inside oases to stabilise the area.

More efficient use of existing water resources and control of salinisation are other effective tools for improving arid lands. New ways are being sought to use surface-water resources such as rain water harvesting or irrigating with seasonal runoff from adjacent highlands. New ways also being sought to find and tap groundwater resources and to develop more effective ways of irrigating arid and semiarid lands.

Research on the reclamation of deserts also is focusing on discovering proper crop rotation to protect the fragile soil, on understanding how sand-fixing plants can be adapted to local environments, and on how grazing lands and water resources can be developed effectively without being overused. Off-road vehicles significantly increase soil loss in the delicate desert environment of the western United States, which can be controlled by restrictions on such vehicles.

Structure of the Soil

Structure is very important since (along with soil texture) it affects the porosity of the soil. A dense structure will greatly reduce the amount of air and water than can move freely through the soil. Also, it will affect the plant's ability to propagate roots through the soil.

There are five major classes of structure seen in soils: platy, prismatic, columnar, granular, and blocky. There are also structureless conditions. Some soils have simple structure, each unit being an entity without component smaller units. Others have compound structure, in which large units are composed of smaller units separated by persistent planes of weakness. These major classes are described below.

Platy Structure: In platy structure, the units are flat and platelike. They are generally oriented horizontally. A special form, lenticular platy structure, is recognised for plates that are thickest in the middle and thin toward the edges. Platy structure is usually found in subsurface soils that have been subject to leaching or compaction by animals or machinery. The plates can be separated with a little effort by prying the horizontal layers with a pen knife. Platy structure tends to impede the downward movement of water and plant roots through the soil.

Prismatic Structure: In prismatic structure, the individual units are bounded by flat to rounded vertical faces. Units are distinctly longer vertically, and the faces are typically casts or molds of adjoining units. Vertices are angular or subrounded; the tops of the prisms are somewhat indistinct and normally flat. Prismatic structures are characteristic of the B horizons or subsoils. The vertical cracks result from freezing and thawing and wetting and drying as well as the downward movement of water and roots.

Columnar Structure: In columnar structure, the units are similar to prisms and are bounded by flat or slightly rounded vertical faces. The tops of columns, in contrast to those of prisms, are very distinct and normally rounded. Columnar structure is common in the subsoil of sodium affected soils. Columnar structure is very dense and it is very difficult for plant roots to penetrate these layers. Techniques such as deep plowing have help to restore some degree of fertility to these soils.

Blocky Structure: In blocky structure, the structural units are blocklike or polyhedral. They are bounded by flat or slightly rounded surfaces that are casts of the faces of surrounding peds. Typically, blocky structural units are nearly equidimensional but grade to prisms and to plates. The structure is described as angular blocky if the faces intersect at relatively sharp angles; as subangular blocky if the faces are a mixture of rounded and plane faces and the corners are mostly rounded.

Blocky structures are common in subsoil but also occur in surface soils that have a high clay content. The strongest blocky structure is formed as a result of swelling and shrinking of the clay minerals which produce cracks. Sometimes the surface of dried-up sloughs and ponds shows characteristic cracking and peeling due to clays.

Granular Structure: In granular structure, the structural units are approximately spherical or polyhedral and are bounded by curved or very irregular faces that are not casts of adjoining peds. In other words, they look like cookie crumbs. Granular structure is common in the surface soils of rich grasslands and highly amended garden soils with high organic matter content.

Soil mineral particles are both separated and bridged by organic matter breakdown products, and soil biota exudates, making the soil easy to work. Cultivation, earthworms, frost action and rodents mix the soil and decreases the size of the peds. This structure allows for good porosity and easy movement of air and water. This combination of ease in tillage, good moisture and air handling capabilities, and good structure for planting and germination, are definitive of the phrase good tilth.

Structureless Soils

Some soils lack structure and are referred to as structureless. In structureless layers or horizons, no units are observable in place or after the soil has been gently disturbed, such as by tapping a spade containing a slice of soil against a hard surface or dropping a large fragment on the ground. When structureless soils are ruptured, soil fragments, single grains, or both result. Structureless soil material may be either single grain or massive. Soil material of single grains lacks structure. In addition, it is loose.

Surface Runoff

Surface runoff is water, from rain, snowmelt, or other sources, that flows over the land surface, and is a major component of the water cycle. Runoff that occurs on surfaces before reaching a channel is also called overland flow. A land area which produces runoff draining to a common point is called a watershed. When runoff flows along the ground, it can pick up soil contaminants such as petroleum, pesticides, or fertilisers that become discharge or overland flow. Surface runoff may be generated in three ways:

1. When the rate of rainfall on a surface exceeds the rate at which water can infiltrate the ground, and the depression storage has been filled. This is called infiltration excess overland flow, Hortonian overland flow (after Robert E. Horton), or unsaturated overland flow. This more commonly occurs in arid and semi-arid regions, where rainfall intensities are high and the soil infiltration capacity is reduced because of surface sealing, or in paved areas.

2. When the soil is saturated and the depression storage filled, and rain continues to fall, the rainfall will immediately produce surface runoff. (Note in the photo to the left the microdepressions are full of water as seen in the lower left of the image.) This is saturation excess overland flow or saturated overland flow.

3. After water infiltres the soil on an up-slope portion of a hill, the water may flow laterally through the soil, and exfiltrate (flow out of the soil) closer to a channel. This is called subsurface return flow or interflow.

As it flows, the amount of runoff may be reduced in a number of ways: a small portion of it may evaporate; water may become temporarily stored in microtopographic depressions; and a portion of it may become run-on, which is the infiltration of runoff as it flows overland. What surface runoff that remains eventually flows into a receiving water body such as a river, lake, estuary or ocean.

Human Impact

Urbanisation increases surface runoff, by creating more impervious surfaces such as pavement and buildings do not allow percolation of the water down through the soil to the aquifer. It is instead forced directly into streams, where erosion and siltation can be major problems, even when flooding is not. Increased runoff reduces groundwater recharge, thus lowering the water table and making droughts worse, especially for farmers and others who depend on water wells.

When anthropogenic contaminants are dissolved or suspended in runoff, the human impact is expanded to create water pollution. This pollutant load can reach various receiving waters such as streams, rivers, lakes, estuaries and oceans with resultant water chemistry changes to these water systems and their related ecosystems.

A contrarian could note there is considerable surface runoff in natural systems from animal wastes being entrained in runoff or from natural sediment loading in the absence of human alteration of the land. While these statements are true, they fail to convey that the most pernicious consequences to human health and ecosystems are from runoff issues related human intervention; however, in underdeveloped countries the proportion of runoff attributable to natural factors has greater dominance, principally due to the lack of isolation of water supplies from potential animal waste carrying runoff.

Effects of Surface Runoff

Soil erosion

Surface runoff is one of the causes of erosion of the earth's surface. Erosion is the displacement of solids (soil, mud, rock and other particles) by the agents of wind, water or ice, by downward or down-slope movement in response to gravity or by living organisms (in the case of bioerosion). Erosion is to be distinguished from weathering, which is the decomposition of rock, although the two processes may be concurrent. Erosion is an intrinsic natural process but in many places it is increased by human land use.

Poor land use practices include deforestation, overgrazing, unmanaged construction activity and road or trail building. However, improved land use practices can limit erosion, using techniques like terrace-building and tree planting. A certain amount of erosion is natural and, in fact, healthy for the ecosystem. For example, gravels continually move downstream in watercourses. Excessive erosion, however, can cause problems, such as receiving water sedimentation, ecosystem damage (including fish kills) and outright loss of soil.

What causes erosion to be severe in some areas and minor elsewhere? It is a combination of many factors, including the amount and intensity of precipitation, the texture of the soil, the gradient of the slope, ground cover (from vegetation, rocks, etc.) and land use. The first factor, rain, is the agent for erosion, but the degree of erosion is governed by other factors. The first three factors can remain fairly constant over time.

In general, given the same kind of vegetative cover, you expect areas with high-intensity precipitation, sandy or silty soils and steep slopes to be the most erosive. Soils with a greater proportion of clay that receive less intense precipitation and are on gentle slopes tend to erode less. The factor that is most subject to change is the amount and type of ground cover. When fires burn an area or when vegetation is removed as part of timber operations or building a house or a road, the susceptibility of the soil to erosion is greatly increased.

Roads are especially likely to cause increased rates of erosion because, in addition to removing ground cover, they can significantly change drainage patterns. A road that has a lot of rock and one that is "hydrologically invisible" (that gets the water off the road as quickly as possible, mimicking natural drainage patterns) has the best chance of not causing increased erosion. Understandably, many human activities remove vegetation from an area, making the soil easily eroded.

Logging and heavy grazing can reduce vegetation enough to increase erosion. Changes in the kind of vegetation in an area can also effect erosion rates. Different kinds of vegetation effect infiltration rates of rain into the soil. Forested areas have higher infiltration rates, so precipitation will result in less surface runoff, which erodes. Instead much of the water will go in subsurface flows, which are generally less erosive. Leaf litter and low shrubs are an important part of the high infiltration rates of forested systems, the removal of which can increase erosion rates.

Leaf litter also shelters the soil from the impact of falling raindrops, which is a significant agent of erosion. Vegetation can also change the speed of surface runoff flows, so grasses and shrubs can also be instrumental in this aspect. One of the main causes of erosive soil loss in the year 2006 is the result of slash and burn treatment of tropical forests. When the total ground surface is stripped of vegetation and then seared of all living organisms, the upper soils are vulnerable to both wind and water erosion.

In a number of regions of the earth, entire sectors of a country have been rendered unproductive. For example, on the Madagascar high central plateau, comprising approximately ten percent of that country's land area, virtually the entire landscape is sterile of vegetation, with gully erosive furrows typically in excess of 50 meters deep and one kilometer wide. Shifting cultivation is a farming system which sometimes incorporates the slash and burn method in some regions of the world.

When land is overused by animal activities (including humans), there can be mechanical erosion and also removal of vegetation leading to erosion. In the case of the animal kingdom, this effect would become material primarily with very large animal herds stampeding such as the Blue Wildebeest on the Serengeti plain. Even in this case there are broader material benefits to the ecosystem, such as continuing the survival of grasslands, that are indigenous to this region. This effect may be viewed as anomalous or a problem only when there is a significant imbalance or overpopulation of one species.

In the case of human use, the effects are also generally linked to overpopulation. For when large numbers of hikers use trails or extensive off road vehicle use occurs, erosive effects often follow, arising from vegetation removal and furrowing of foot traffic and off road vehicle tires. These effects can also accumulate from a variety of outdoor human activities, again simply arising from too many people using a finite land resource.

One of the most serious and long-running water erosion problems worldwide is in China, on the middle reaches of the Yellow River and the upper reaches of the

Yangtze River. From the Yellow River, over 1.6 billion tons of sediment flow each year into the ocean. The sediment originates primarily from water erosion in the Loess Plateau region of northwest China.

Gravity erosion: Mass wasting is the down-slope movement of rock and sediments, mainly due to the force of gravity. Mass wasting is an important part of the erosional process, as it moves material from higher elevations to lower elevations where transporting agents like streams and glaciers can then pick up the material and move it to even lower elevations. Mass-wasting processes are occurring continuously on all slopes; some mass-wasting processes act very slowly, others occur very suddenly, often with disastrous results.

Any perceptible down-slope movement of rock or sediment is often referred to in general terms as a landslide. However, landslides can be classified in a much more detailed way that reflects the mechanisms responsible for the movement and the velocity at which the movement occurs. One of the visible topographical manifestations of a very slow form of such activity is a scree slope. Slumping happens on steep hillsides, occurring along distinct fracture zones, often within materials like clay that, once released, may move quite rapidly downhill. They will often show a spoon-shaped depression, within which the material has begun to slide downhill.

In some cases, the slump is caused by water beneath the slope weakening it. In many cases it is simply the result of poor engineering along highways where it is a regular occurrence. Surface creep is the slow movement of soil and rock debris by gravity which is usually not perceptible except through extended observation. However, the term can also describe the rolling of dislodged soil particles 0.5 to 1.0 mm in diameter by wind along the soil surface.

Water erosion: Splash erosion is the detachment and airborne movement of small soil particles caused by the impact of raindrops on soil. Sheet erosion is the result of heavy rain on bare soil where water flows as a sheet down any gradient, carrying soil particles. Where precipitation rates exceed soil infiltration rates, runoff occurs. Surface runoff turbulence can often cause more erosion than the initial raindrop impact. Gully erosion results where water flows along a linear depression eroding a trench or gully. Valley or stream erosion occurs with continued water flow along a linear feature.

The erosion is both downward, deepening the valley, and headward, extending the valley into the hillside. In the earliest stage of stream erosion, the erosive activity is dominantly vertical, the valleys have a typical V cross-section and the stream gradient is relatively steep. When some base level is reached the erosive activity switches to lateral erosion which widens the valley floor and creates a narrow floodplain. The stream gradient becomes nearly flat and lateral deposition of sediments becomes important as the stream meanders across the valley floor.

In all stages of stream erosion by far the most erosion occurs during times of flood, when more and faster-moving water is available to carry a larger sediment load. In such processes, it is not the water alone that erodes, suspended abrasive particles, pebbles and boulders can also act erosively, as they traverse a surface.

Shoreline erosion: Shoreline erosion, on both exposed and sheltered coasts, primarily occurs through the action of currents and waves but sea level (tidal) change can also play a role. Sediment is transported along the coast in the direction of the prevailing current (longshore drift).

When the upcurrent amount of sediment is less than the amount being carried away, erosion occurs. When the upcurrent amount of sediment is greater, sand or gravel banks will tend to form. These banks may slowly migrate along the coast in the direction of the longshore drift, alternately protecting and exposing parts of the coastline.

Ice erosion: Ice erosion is caused by movement of ice, typically as glaciers. Glaciers can scrape down a slope and break up rock and then transport it, leaving moraines, drumlins and glacial erratics in their wake, typically at the terminus or during glacier retreat. Ice wedging is the weathering process in which water trapped in tiny rock cracks freezes and expands, breaking the rock. This can lead to gravity erosion on steep slopes. The scree which forms at the bottom of a steep mountainside is mostly formed from pieces of rock broken away by this means. It is a common engineering problem, wherever rock cliffs are alongside roads, because morning thaws can drop hazardous rock pieces onto the road.

Wind erosion: Wind erosion, also known as eolian erosion is the movement of rock and/or sediment by the wind. Windbreaks are often planted by farmers to reduce wind erosion. This includes the planting of trees, shrubs, or other vegetation, usually perpendicular or nearly so to the principal wind direction. The wind causes dust particles to be lifted and therefore moved to another region. Wind erosion generally occurs in areas with little or no vegetation, often in areas where there is insufficient rainfall to support vegetation. An example is the formation of sand dunes, on a beach or in a desert.

Tectonic effects: The removal by erosion of large amounts of rock, from a particular region and its deposition elsewhere, can result in a lightening of the load on the lower crust and mantle. This can cause tectonic or isostatic uplift in the region.

Influence of environmental

The principal environmental issues associated with runoff are the impacts to surface water, groundwater and soil through transport of water pollutants to these systems. Ultimately these consequences translate into human health risk, ecosystem distubance and aesthetic impact to water resources. Some of the contaminants that create the greatest impact to surface waters arising from runoff are petroleum substances, herbicides and fertilisers. Quantitative uptake by surface runoff of pesticides and other contaminants has been studied since the 1960s, and early on contact of pesticides with water was known to enhance phytotoxicity.

In the case of surface waters, the impacts translate to water pollution, since the streams and rivers have received runoff carrying various chemicals or sediments. When surface waters are used as potable water supplies, they can be compromised regarding health risks and drinking water aesthetics (that is, odor, color and turbidity effects). Contaminated surface waters risk altering the metabolic processes of the aquatic species that they host; these alterations can lead to death, such as fish kills, or alter the balance of populations present.

Other specific impacts are on animal mating, spawning, egg and larvae viability, juvenile survival and plant productivity. In the case of groundwater, the main issue is contamination of drinking water, if the aquifer is abstracted for human use. Regarding soil contamination, runoff waters can have two important pathways of concern. Firstly, runoff water can extract soil contaminants and carry them in the form of water pollution to even more sensitive aquatic habitats. Secondly, runoff can deposit contaminants on relatively pristine soils, creating health or ecological consequences.

Flooding

Flooding occurs when a watercourse is unable to convey the quantity of runoff flowing downstream. The frequency with which this occurs is described by a return period. Flooding is a natural process, which maintains ecosystem composition and processes, but it can also be altered by land use changes such as river engineering. Floods can be both beneficial to societies or cause damage. Agriculture along the Nile floodplain took advantage of the seasonal flooding that deposited nutrients beneficial for crops.

However, as the number and susceptability of settlements increase, flooding increasingly becomes a natural hazard. Adverse impacts span loss of life, property damage, contamination of water supplies, loss of crops, and social dislocation and temporary homelessness. Floods are among the most devastating of natural disasters.

Agricultural Issues

A common context of run-off deals with agriculture. When farmland is tilled and bare soil is revealed, rainwater carries billions of tons of topsoil into waterways each year, causing loss of valuable topsoil and adding sediment to produce turbidity in surface waters. The other context of agricultural issues involves the transport of agricultural chemicals (nitrates, phosphates, pesticides, herbicides etc) via surface runoff.

This result occurs when chemical use is excessive or poorly timed with respect to high precipitation. The resulting contaminated runoff represents not only a waste of agricultural chemicals, but also an environmental threat to downstream ecosystems. The alternative to conventional farming is organic farming which eliminates or greatly reduces chemical usage.

Analysis of Runoff

Runoff is analysed by using mathematical models in combination with various water quality sampling methods. Measurements can be made using continuous automated water quality analysis instruments targeted on pollutants such as specific organic or inorganic chemicals, pH, turbidity etc. or targeted on secondary indicators such as dissolved oxygen. Measurements can also be made in batch form by extracting a single water sample and conducting any number of chemical or physical tests on that sample. In the 1950s or earlier hydrology transport models appeared to calculate quantities of runoff, primarily for flood forecasting.

Beginning in the early 1970s computer models were developed to analyse the transport of runoff carrying water pollutants, which considered dissolution rates of various chemicals, infiltration into soils and ultimate pollutant load delivered to receiving waters. One of the earliest models addressing chemical dissolution in runoff and resulting transport was developed in the early 1970s under contract to the EPA. This computer model formed the basis of much of the mitigation study that led to strategies for land use and chemical handling controls. Other computer models have been developed (such as the DSSAM Model) that allow surface runoff to be tracked through a river course as reactive water pollutants. In this case the surface runoff may be considered to be a line source of water pollution to the receiving waters.

Mitigation of adverse impacts of runoff can take several forms:

- Land use development controls aimed at minimising impervious surfaces in urban areas.
- Erosion contols for farms and construction sites.
- Flood control programmes.
- Chemical use and handling controls in agriculture, landscape maintenance, industrial use etc.

Regarding Land use controls, the U.S. Environmental Protection Agency and others have encouraged research on methods of mimimising total surface runoff by avoiding unnecessary hardscape. Many municpalities have produced guidelines and codes for land developers that encourage minimum width sidewalks, use of pavers set in earth for driveways and walkways and other design techniques to allow maximum water infiltration in urban settings. An example land use control programme can be viewed at seen in the city of Santa Monica, California. Erosion controls have appeared since medieval times when farmers realised the importance of contour farming to protect soil resources. Beginning in the 1950s these agricultural methods became increasingly more sophisticated.

After passage of the National Environmental Policy Act (NEPA) in the United States, much more effort was focussed upon mitigation of construction runoff by such tactics as: use of hay bales and barriers to slow runoff on slopes, programming construction for months that have less rainfall and mimimising extent and duration of exposed graded areas. Flood control programmes as early as the first half of the twentieth century became quantitative in predicting peak flows of riverine systems. Progressively strategies have been developed to minimise peak flows and also to reduce channel velocities. Some of the techniques commonly applied are: provision of holding ponds to buffer riverine peak flows, use of energy dissipators in channels to reduce stream velocity and land use controls to minimise runoff.

Chemical use and handling has become a focal point mainly since passage of NEPA in the U.S. States and cities have become more vigilant in controlling the containment and storage of toxic chemicals, thus preventing releases and leakage. Methods commonly applied are: requirements for double containment of underground storage tanks, registration of hazardous materials usage, reduction in numbers of allowed pesticides and more stringent regulation of fertilisers and herbicides in landscape maintenance. In many industrial cases, pretreatment of wastes is required, to minimise escape of pollutants into sanitary or stormwater sewers.

8

Soil Moisture for Plant Production

In many parts of the subhumid and semiarid tropics, crop yields are declining on response to inputs such as fertilisers, and droughts and shortages of irrigation water are increasingly evident. The problem of lower annual increases in yield per hectare is a not confined to irrigated rice. Wheat and maize are also apparently reaching similar plateaus.

For the three major staple crops (paddy rice, wheat and maize) the average yield increases between 1963 and 1983 were respectively, 2.1, 3.6 and 2.9 percent per year; but in the ten years 1983-1993 the rates of increase had fallen respectively to 1.5, 2.1 and 2.5 percent per year. Some 40 percent of all food is produced under irrigation, from about 18 percent of the world's area of arable land plus permanent crops, with 60 percent produced under rainfed agriculture. As populations have risen this arable land and permanent crops area per head has been falling, except in the case of Europe.

In the State of Paraná in southern Brazil, from the time of clearing the land from native forest decades ago, yields of crops under conventional tillage fell between 5 to 15 percent in ten years. This was accompanied by severe losses of soil, associated organic matter and applied nutrients, resulting in downstream flooding, sedimentation and other damage. Records from Lesotho show that mean yields of major crops (generally without added fertilisers) declined between 1978 and 1986, related to a combination of adverse weather conditions, decline in soil conditions, and recurrent erosion and runoff. A three-year running mean of yields has been used, in order to smooth the effects of between-year weather variations.

At the same time, it has been widely observed that ongoing land degradation across topographic catchments has resulted in increasingly irregular streamflow, with more floods of muddy water in the rainy season and declining volume and duration of streamflow during the dry season. Human-induced agricultural land degradation is widespread in irrigated and rainfed land and in both tropical and temperate zones. Land degradation represents a challenge to the sustainability of farming systems in all regions, even those of low population densities. In rainfed lands, compaction, erosion and runoff are significant problems. On irrigated lands, problems are often those of poor drainage control, salinisation and compaction leading to nutrient deficiencies.

Such problems are not confined to tropical areas. Clear-felling of trees, grazing on very steep slopes and the compacting effects of farm machinery result in excess water runoff and erosion in temperate zones. Estimates of damage caused by compaction and erosion in the Eurasian region suggest that about 327 million hectares of land in Eurasia have been severely affected by wind and water erosion. Approximately 170 million ha of land have been affected by soil compaction. Conservative estimates calculate a production loss of 15 million tons of grain, two million tons of sugar beet and 500 000 tons of maize.

Others calculate a 16-27 percent decrease in production as a result of compaction, with a loss of 50 million tons of grain production alone.

A study of the effects of soil compaction on wheat production in New Zealand showed that as the soil becomes increasingly degraded, costs rise as yields fall, squeezing margins of profit per hectare. This indicates a wider problem that, as yields begin to decline, farmers may apply more fertiliser, masking the underlying decline into unsustainable and uneconomic production. A survey of small resource-poor farmers in central Paraguay showed that as erosion and runoff continued, yields of cotton, tobacco, maize and other crops declined. As a result net farm incomes fell and farmers could no longer afford to buy equipment or inputs that might help to reverse the downward trend. This has led to farms being abandoned and desperate families migrating to the cities in search of income that farming could no longer provide.

Potential sources of growth in overall output are:

- expansion of arable land area,
- increases in cropping intensities to give greater harvested area, and
- growth in yield per hectare.

Expansion of arable land will be limited because almost all lands of good and moderate quality have been settled. Expansion of the cultivated area will be onto land with increasing difficulties and hazards, which will reflect negatively on the yields and economics in crop production of both rainfed and irrigated crops. Increases in cropping intensity with shorter (or even no) regular recuperative periods during which damaged soils can recover their soil fertility will result in continued and worsening land degradation. Rates of yield increase are tending to fall and upper limits of potential yield, at least of key grain crops, are apparently being approached where high inputs are used on the best soils.

Intensification is the only option to increase usable biomass and available water per unit area of land. The challenge is to achieve intensification without causing more damage to soils and to the quantity and reliability of water supplies. Unfortunately, past attempts to intensify production using conventional methods have often resulted in damage to the soil. The sustainability of agriculture depends not only on the soil continuing to be a fit place for crops, pastures and trees, but also on young people being enthused by farming to provide a continuity from one generation to the next, developing and carrying forward up-to-date knowledge and relevant skills in the husbandry of plants, animals and land.

How can farming return to being a way of life which is satisfying to many and that encourages them to remain in the rural areas? How can sufficient water be ensured, both for plant growth and for the regular flow of rivers, when much recent experience shows increasingly severe effects of drought on crop plants and a decline in the regularity and volume of river flow? How to achieve not only greater total output, but also better quality and improved food security over the year? How to produce a greater variety of foods to improve nutrition and health and reduce poverty by generating income? Conventional approaches seem to be inadequate for the task, despite the efforts of many to date. Key parts of any strategy to address these issues include:

- Recognising the soil as a key and living component of the environment. To date it has received far less attention in comparison with the above-ground components, which are more readily perceived.

- Encouraging the inherent capacities of life itself. Particularly the ability of bacteria, fungi, soil fauna and plants to continually colonise and modify habitats.
- Prolonging the usefulness of rainwater and organic matter. By recycling through different biotic processes as many times as possible.

Hydrological Cycle

An understanding of the hydrological cycle is essential for the effective management of rainwater and soil water. Water occurs not only as a liquid, but also as a solid (*e.g.* hail, snow) and as a gas—water vapour. The total amount of water in the world is constant, but water is continuously changing from one form to another and is continuously moving at different speeds. These interrelationships are shown in a simplified form at a regional scale in Figure 1.

Heat from the sun causes water at the surface of oceans, lakes and rivers to change into water vapour in a process called evaporation. Transpiration in plants is a similar process, in which water is absorbed from the soil by plant roots and transported up the stem to the leaves, from where it is released (transpired) as water vapour into the atmosphere. As the water vapour produced by evaporation and transpiration rises into the atmosphere, so the temperature decreases and water vapour changes into water droplets (condensation), which accumulate as clouds. Depending on their size, these may be released as rainfall.

Once rainfall reaches the land surface it can infiltrate into the soil, run off over the surface as overland flow, or accumulate on plant leaves or in puddles from where it evaporates back to the atmosphere. A combination of these processes is commonly the case. The rainfall that infiltrates into the soil forms part of the soil water, of which some may be used by plants for transpiration, some may return to the atmosphere through evaporation from the soil surface, and some—if sufficient infiltration occurs—may move beyond the rooting zone to the groundwater.

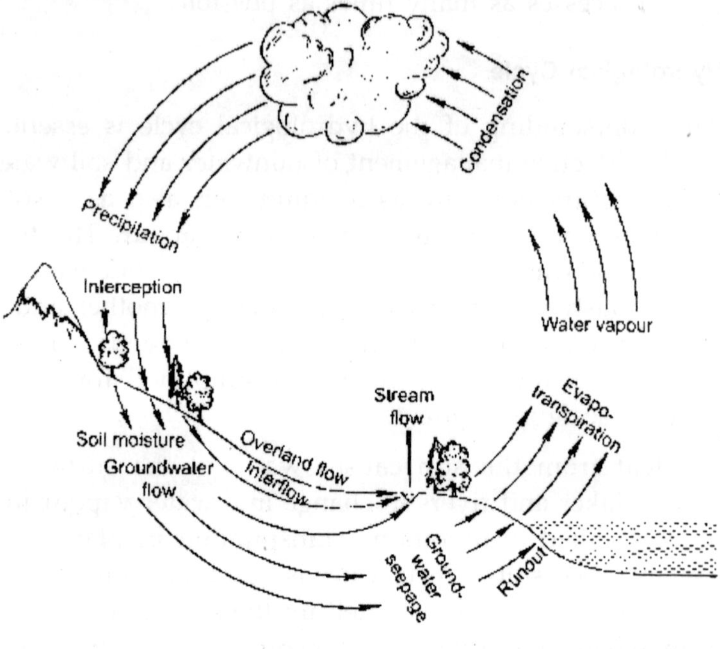

Figure 1. Simplified diagram of the hydrological cycle

The groundwater moves laterally and slowly towards the sea to complete the hydrological cycle, but part of it will seep into springs, streams, rivers and lakes on the way. In this way the groundwater maintains the water level in wells, and the continuity of river and streamflow during dry periods. Rainwater that runs off the land moves rapidly downhill towards river courses, contributing to peak flows, and is of great concern.

Runoff is not only a waste of rainfall that could have contributed to crop production and groundwater supplies, but it frequently causes floods or damage to roads and farmland, and erodes soil that is redeposited in river courses and reservoirs downstream. Groundwater is derived from rainwater that has infiltrated into the soil and drained beyond the rooting zone in excess of both the quantity needed for the crop or the vegetation and the water-storage capacity of the soil. Groundwater moves very slowly through subsoil materials in the direction of the dominant drainage course.

If its upper surface, the water table, does not sink below the level of the streambed, water is released to springs that feed streams and tributaries. This occurs throughout the year and in this way groundwater acts as a buffer in maintaining stream base flows and water levels in wells during dry periods. In soils with relatively impermeable subsoil layers beneath more highly permeable layers, perched water tables may develop above the groundwater, due to water being held up by the impermeable layers. The water in a perched water table, sometimes referred to as interflow, will slowly move laterally and may emerge into stream courses or springs at lower elevations.

It does not contribute directly to the groundwater. The presence of groundwater or a perched water table is indicated by saturated soils, and usually by a dominance of light grey, bluish-grey, bluish or greenish colours. These colours are typical of certain iron compounds that only form in waterlogged soils where oxygen is lacking. The amount of rainfall that percolates beyond the lower limit of the rooting zone towards the groundwater will depend on the amount of water used for transpiration by the crops or vegetation.

For a particular climate and soil type, forest transpires more water than grassland, which generally uses more water than crops. The high water use by forest is due to its generally greater transpiration rate, the longer period of transpiration compared with crops, and the deeper roots enabling it to absorb water from greater depths. Changes in land use can therefore affect the quantity of water transpired and hence the quantity reaching the groundwater. Replacing forest vegetation with grassland or annual crops may increase deep drainage and so provide higher base flows in streams and rivers. Changes in soil management can also affect the quantity of deep drainage replenishing groundwater.

The introduction of poor management practices that increase the proportion of rainfall lost as runoff will reduce base flows and increase peak flows and the incidence of flooding. Conversely, an improvement in soil and nutrient management will lead to higher grain and foliage production, higher transpiration rates, and hence less recharge. In order to consider rainwater for plants and for groundwater as parts of a sequence, it is important to have a mental picture of its journey. After passing through the atmosphere in response to gravity, water from rain or irrigation travels to some or all of the following destinations.

Management of the soil can significantly affect runoff; direct evaporation from the soil surface; the amount of soil moisture available to plants within range of their roots; and the depth to which roots can penetrate. How much water reaches each of these destinations over a given period depends on the physical condition of the soil and its influence on infiltration and runoff, and on the atmospheric conditions as they affect evaporation and transpiration.

Catchments

Water caught by a catchment will flow towards the lowest point at the outlet, where it may join water emerging from other catchments. The outer boundaries of a catchment are defined by ridgelines along the crests of the surrounding uplands. From the sides of a valley surface runoff tends to flow perpendicularly to the slope from crest to streamline. A watershed is the area of land dividing two streamlines. Water moves away from the crest line towards the streamlines on either side. Thus a hill slope can be considered as either the inner slope of a catchment or the outer slope of the watershed.

Underlying geological formations, together with weathering and uplift processes, affect the form of landscapes. They influence the steepness or shallowness of slopes, whether the streamlines are of relatively sinuous shape or with abrupt changes of direction. The flow of water along the streamlines tends to cut the heads of streamlines back into the underlying materials. For the purposes of enabling rainwater to soak into the soil and controlling the rate of flow of any excess runoff, we can subdivide a given catchment into a more detailed hierarchy of catchments, in which the smallest subdivisions may be measured in square centimetres, the larger in hectares, within catchments of square kilometres. Rainfall entry into the soil depends on the porosity of the soil at any scale, while management of runoff and erosion across the surface also depends on any physical works that may be constructed when rainfall rates exceed even the best infiltration rate.

A stream catchment may be large or small, of steep or shallow slope and composed of natural subcatchments and then of field catchments. In cropping agriculture the next smaller subdivision is the bund catchment, between any pair of physical conservation banks with its ridges and furrows (formal or informal) along planting rows. The physical conservation banks, vegetated with a fodder grass, subdivide the field catchment as well as separating the bund catchments. The aim of these structures is to conserve water and soil, though their effects on yield are disappointing. Their most significant function is to provide guidelines for contour planting of crop rows.

The smallest subdivision to trap rainwater and give it time to soak in is the microcatchment with its mulch. In forestry, because the young trees are more widely spaced, the same effect can be obtained by a set of half moon shaped microcatchments, one at each planting position. Within this overall framework, the key to infiltration is to keep the soil porous with a cover of crop residues, which prevents damaging raindrop impact and provides a substrate for soil organisms.

The conservation effects of forests are due not so much to the presence of the trees themselves but to the litter of fallen leaves, twigs and branches, plus any low-growing vegetation. If the soil surface has not been damaged by trampling, less rainwater will run off and more will infiltrate into the soil. How much plant-available soil moisture remains at a given time depends on the texture and porosity of the soil, the previous volume of soil moisture, the volume removed by direct evaporation, evapotranspiration and deep drainage. Irrigation (if available) is normally required when about two thirds of the available water—between field capacity (FC) and permanent wilting point (PWP)—has been depleted.

If irrigation is not an option, it makes sense to manage the soil to develop and retain a maximum amount of soil pores of a wide range of sizes. This will maximise the capacity for water retention and enable plants to withstand drought for longer periods. Loam textures generally have the largest available water capacity, while sand on the one extreme has a small available water capacity, as does clay at the other.

Available water capacity coupled with soil depth determines the volume of water usable by plants at a particular site. This is illustrated by comparing relevant characteristics and consequent amounts of available water for two soils on which tea is grown, one at Timbilil in Kenya, the other at Marikitanda in Tanzania. Flood flows in streams and rivers, which rise quickly after heavy rainfall derive mostly from rapid overland flow of water. Flood flows are often muddy with eroded materials. Clear streamflow originates from rainwater, which has infiltrated the soil and percolated through pores of a range of sizes at different slower speeds.

As dug wells provide direct access to shallow groundwater, on which many rural communities rely, it is important that enough rainwater penetrates and pass through the soil to replenish groundwater. Exceptional rainfall such as during typhoons or hurricanes on already saturated soil can result in floods and erosion that change the landscape, no matter how well the land and the crops, grassland or forest are managed.

Pore Spaces in Soils

Although we generally think of soil in terms of its solid parts, *i.e.* the sand, silt, clay and organic matter, it is the spaces between these solid particles that are as important as the solid particles. This is because it is the spaces where all the action takes place, just as in a house, where all the important activities occur in the rooms rather than in the walls and floors. It is therefore the architecture of the soil that is important. The pore spaces in a soil vary in abundance according to the type of soil and how it has been managed.

Soils under natural vegetation generally exhibit high porosity because of high biological activity and lack of interference by man. Consequently they have superior physical qualities compared with most soils used for crops or grazing. Pore spaces in soils vary in size, and both the size and continuity of pores have an important influence on the types of activities that occur in soil pores.

Pore sizes from 0.0002 to 0.05 mm diameter retain water that can be absorbed by crops and are referred to as storage pores, whereas smaller pores (the residual pores) hold water too tightly for plants to be able to extract it. Pores larger than about 0.05 mm diameter, referred to as transmission pores, allow water to drain through the soil and enable air to enter the pores as the water drains out. Pore spaces are also needed for roots to freely penetrate soils in order to take up nutrients and water.

The sizes of roots vary with the type of crop, but the smallest roots, apart from root hairs, have diameters of 0.1 to 0.3 mm and so soils must have pore spaces of at least this size if the smaller roots are to penetrate freely. In most soils roots grow partly through existing pores, the transmission pores, and partly by moving aside soil particles. Roots can only force their way into smaller pores if the soils are sufficiently compressible; the compressibility of soils increases with increasing water content, since water provides a form of lubrication between soil particles.

Water Movement in Soil

The amount of water present in a soil, which is available

for crop production, will depend on how much of the rainwater remains in the soil after the losses by runoff, evaporation, and deep drainage. The amount of rainfall that reaches the groundwater and thus contributes to water security, will depend on the extent to which the rainfall infiltrating the soil is in excess of that needed to replenish the soil's water holding capacity and satisfy the transpiration needs of the crops.

Good rainwater management aims to maximise the amount of rainwater that enters the soil, and to make best use of it while it is there for use by crops and for recharging the groundwater. Any truly unavoidable surface runoff is conducted away safely in such a manner that it does not cause erosion problems. When a well drained soil is saturated to the limit of its rooting zone, the rainwater that does not drain out of the root zone within 48 hours will be retained in soil pores smaller than about 0.05 mm diameter (the critical pore size may vary from 0.03 to 0.1 mm diameter).

The quantity of water retained after 48 hours corresponds to the soil's field capacity (FC). The forces (or suctions) with which this water is held will vary according to pore size. The largest pores still to retain water will hold the water at about a tenth to a third of the pressure of the atmosphere, depending on what suction corresponds to the soil's FC; this will vary with soil type and depth of the water table. The maximum suction most crops can exert to withdraw water from soil varies with the crop, but the generally accepted value is equivalent to about 15 times the pressure of the atmosphere (*i.e.* 1.5 Mpa). This is approximately equivalent to the pressure that would be experienced when supporting a tonne weight on the palm of the hand.

When soil water has been exhausted down to fifteen bars, the water remaining in the soil will be that stored in pores smaller than 0.0002 mm diameter, and will correspond to the soil's PWP. Water held at suctions greater than the PWP is not available for plant growth. Consequently, it is water held between FC and PWP which can be used by crops for transpiration, and is termed the soil's Available Water Capacity (AWC). However, after a heavy rainstorm some of the water in excess of the soil's FC may be used by a crop while this excess water is percolating through the rooting zone.

The available water held within the range FC to PWP is retained with different strength, and about a third of it is not easily or rapidly available to crops, especially if the crops are transpiring strongly. The higher the transpiration demand, the more available (*i.e.* the less strongly held) the soil water must be to avoid crop water stress. In contrast, for a slowly transpiring crop even water held at higher suctions can be used without causing stress. The maximum amount of available water that a soil can retain (*i.e.* the available water capacity) will vary with the soil's texture, organic matter content, rooting depth and structure.

Soil organic matter is particularly important in that it can retain about twenty times its weight of water. Organic soils and medium textured loamy soils with high contents of very fine sand and silt generally have the highest AWCs, clayey soils intermediate values, and soils with high contents of coarse sand the lowest AWCs. The stone content of soils can also be very important depending on the nature and abundance of the stones. Some ironstone gravel > 2 mm diameter can contain more than 20 percent water (m^3/m^3) at FC and porous limestone and chalk can also make significant contributions to the AWC of a soil. In contrast, a high content of non-porous stones will greatly diminish the AWC of a soil.

For any given soil, the greater the rooting depth, the larger will be the quantity of soil water available to the crop. This is particularly important for annual crops as they have less time to develop deep and extensive rooting systems than perennial crops. The available water capacity may influence the length of growing period for crops grown on that soil. Soils of high available water capacity will permit longer growing periods because of their ability to provide greater quantities of stored water during dry periods than soils of low available water capacity. Shallow soils have little available water, and even in wet years they will be unable to benefit by storing any more water.

Infiltration of Rainwater

In most areas where water shortages occur, maximising the infiltration of rainfall into soil is indispensable to achieving food and water security. Land management should encourage infiltration as opposed to runoff. Exceptions are where rainwater harvesting is necessary for crop production and where high infiltration can lead to risks of landslides or other forms of mass movement. The amount of rainfall that infiltrates will be governed by the intensity of the rainstorm in relation to the soil's infiltration rate.

Excessive tillage and loss of soil organic mater often result in reduced infiltration rate due to loss of surface porosity. When storm intensity is greater than soil infiltration rate, runoff will occur, resulting in a waste of water that should have been used for crop production and for recharging the groundwater. The rate at which rainfall infiltrates into soil is influenced by the abundance, stability and size of the pores at the soil surface, their water content and by the continuity of the transmission pores into the rooting zone. In many soils the number of surface pores is rapidly reduced by the impact of raindrops, which break surface soil aggregates into small particles that clog surface pores and form surface seals or crusts with very few pores.

The destructive raindrop action is avoided where there is a protective cover of crop foliage, residues, mulches or even weeds at or over the soil surface. Other factors that can reduce the number, proportion and continuity of transmission pores are traffic by machinery, humans and animals, which destroys large pores by compaction, and tillage which disrupts the continuity of transmission pores through the smearing and compression of pores during plough pan formation in the subsoil. Infiltration rates are also affected by

- the quantity of water present in the soil at the time of the rainstorm, which will depend on when the last rainstorm occurred and the permeability of the soil, and
- the soil's capacity to retain water, which will vary with soil depth, stoniness, and texture.

Percolation of Rainwater

When a heavy rainstorm falls on a well-structured soil, rainwater percolates down through the dry soil as a wetting front, temporarily saturating the soil and displacing air. This is accompanied by the rapid drainage of water from the larger pores (greater than 0.05 mm) through gravity and the pressure of the mass of rainwater above. These larger pores exert only small forces of attraction on soil water. After about two days of drainage field capacity will have been attained and air will have re-entered the larger pores. In poorly structured soils, rainwater will drain much more slowly.

Drainage often continues for several weeks depending on the depth to the slowest horizon and the continuity of the larger pores with depth. In fine textured soils with cracks drainage water will flow down through the cracks in heavy rainstorms before the soil is saturated and while parts of the soil profile may still be dry. If the drainage water subsequently enters a smaller pore while passing through the soil, it will be retained, otherwise it will continue until it reaches the water table and contributes to the recharge of groundwater.

Once the drainage water has been lost from the rooting zone, further water movement within the root zone is slow and is referred to as capillary movement. This movement is caused by forces of attraction, known as surface tension forces, which are exerted by soil particles on water. This movement can occur in any direction and includes the upward movement of water from water tables. Surface tension forces pull water into pores within the soil and the smaller the pores the more strongly the water is attracted and held.

Loss of Water

Water is also able to move through soils as water vapour. The most important example of this is the loss of water vapour by evaporation from soil surfaces. This occurs when the concentration of water vapour in the soil close to the surface is higher than that in the atmosphere immediately above. Water vapour will then move from the soil into the atmosphere. The drier and hotter the atmosphere compared with the surface soil, the greater will be the rate of evaporation from the soil, provided sufficient water can be supplied to the surface by capillary movement from below. Fine textured soils have an abundance of small pores and so more capillary movement of water to the surface will generally occur in fine textured than in coarse textured soils.

Water Movements into Plant

Crops use large quantities of water, which under rainfed conditions come entirely from water in the soil, which in turn is derived from rainfall that has infiltrated the soil. A maize crop may use 400 to 750 mm of water depending on the rainfall and evaporation conditions. This corresponds to 4,000 to 7,500 cubic metres of water per hectare over the growing season. Almost all the water absorbed from the soil by crop roots passes up through the stem into the leaves, where it evaporates and passes into the atmosphere in a process known as transpiration. This process accounts for almost all of the water absorbed by plant roots (about 99 percent, the remaining 1 percent being used directly in cell processes).

Transpiration is essentially the same as the process of evaporation. Evaporation is what happens when a bowl of water is left in the sun. The liquid water disappears as it becomes converted into water vapour, and the higher the temperature, the drier the air, and the greater the wind speed, the greater will be the rate of evaporation. Evaporation occurs whenever water is exposed to the atmosphere, *i.e.* from lakes, rivers and puddles, and from the raindrops that accumulate on a leaf after a rainstorm. To ensure an efficient uptake of sufficient water by crops it is important that the crop roots are well distributed and able to penetrate deeply into the soil.

As a soil dries out from the surface downwards, so roots in the deeper layers tend to increase in number to compensate. When soil water reaches the surface of a root or root hair, it moves across the root into the xylem, which contains narrow tubes running through the root and extending up through the stems into the leaves. On reaching the leaves, water passes from the xylem into leaf cells where it evaporates into air spaces within the leaf. These air spaces are saturated with water vapour, and are connected to the normally drier outside air by very small openings in the leaves called stomata.

During the day the stomata open, which allows carbon dioxide to enter the leaf. Sunlight is used to make sugars within the plant: a process known as photosynthesis. Part of the sugars are used to produce energy by a process called respiration, part are converted into substances forming the various plant organs. Photosynthesis occurs only during daylight, whereas respiration occurs all the time. When the stomata open to allow carbon dioxide to enter, water vapour escapes into the drier air outside. For transpiration to occur there must be a continuous supply and movement of water from the soil to the plant to the atmosphere.

The driving force responsible for this movement is the same as for evaporation, and can be simply described as the tendency for water to move, either as a liquid or a vapour, from where it is more abundant to where it is less abundant. In transpiration, water vapour moves from the very humid (*i.e.* high water vapour content) air spaces within the leaf into the drier atmosphere outside the leaf where the water vapour concentration is lower. The movement of water vapour out of the leaf creates a suction (or "pull") on the water in the leaf cells, the xylem, the roots and the soil, so water moves into the root, up the xylem and into the leaves, to replace that which has been lost from the leaves.

In addition to the transpirational suction, which causes water to move from the soil into the root, there is another force attracting water into the root known as osmosis. In osmosis, water moves from where it is more pure to where it is less pure across a semi-permeable membrane. A semi-permeable membrane is a very thin skin, which has pores large enough for water to pass through into the root but not large enough for dissolved salts to pass out of the root. Water therefore passes from the soil where the water is more pure (*i.e.* contains few dissolved salts) across the root surface (a semi-permeable membrane) into the root where the water is less pure (*i.e.* contains more dissolved salts).

Water Stress

Many areas with low and erratic rainfall where crop water stress is common are also deficient in nutrients, and the lack of nutrients is frequently the second most limiting soil factor. An interaction often occurs between soil water and nutrients, which means that soil water can influence the availability of nutrients, and the availability of nutrients can influence the uptake of soil water and a crop's resistance to drought. Thus, both factors can influence each other. Plants contain a certain amount of water within them, which acts as a buffer against times of water shortage, but the amount is too small to last long.

In contrast, plants store sufficient quantities of nutrients within their tissues to provide a buffer for longer periods when nutrients are not being absorbed. Consequently, water deficiencies become more quickly apparent and damaging than nutrient shortages. This suggests that conserving water may often be of prior and quicker benefit than attempting to conserve soil particles per se. In addition, a lack of water also reduces the uptake of nutrients by a crop. This is largely because nutrients can only move to roots through water films within the soil, and so there must be continuous water films connecting the nutrients with the roots.

A lack of soil water continuity, due to drought for example, will severely reduce the rate of nutrient uptake by crops. A lack of soil water will also diminish nutrient availability by reducing microbial activity, which is responsible for the liberation of nitrogen, phosphorus and sulphur from soil organic matter. When there is a drought it is the surface soil, (which generally contains the bulk of the plants' roots and the soil's nutrients) which dries out first, and so while a crop may still be able to absorb water from the subsoil, it may suffer from a lack of nutrients.

A lack of available nutrients in the soil can restrict crop water uptake, especially when nutrients are limiting root development. This occurs most often in soils that are deficient in phosphorus. Applying P fertiliser to P-deficient soils will often promote root development, and as a result crop water uptake. Consequently the beneficial effects of applying P fertilisers are often relatively greater in seasons of lower rainfall than in those of higher rainfall. The effects of drought and nutrient availability on crop yields are difficult to predict, because the effect will depend on when the water or nutrient shortage occurs in relation to the crop's stage of growth, and its needs and sensitivity to a lack of water or nutrients at that time. It is therefore often difficult to assess which factor, *e.g.* water or nutrients, is the more limiting to yield.

The most limiting factor can vary from season to season depending, for example, on when water shortages occur, and even during a season there will probably be periods when water is the main limiting factor, and other periods when nutrients are most limiting. Water shortages often affect whether or not there is a response to fertilisers, and how much fertiliser should be applied. This is particularly common with N fertiliser, where the optimum response is frequently higher in good seasons than in poor seasons.

For example, when there is no shortage of soil water an application of 40 kg/ha may prove to be the optimum application rate, but when water is lacking only 20 kg/ha may be the optimum amount to apply. This creates difficulties in rainfed agriculture: since it is not possible to reliably predict the distribution and amount of rainfall, farmers cannot know how much fertiliser to apply. One approach that can help to overcome this problem is to apply a modest amount of N fertiliser at the beginning of the season assuming low rainfall, and then to apply additional quantities N later if the season appears promising.

Restricted Rooting

The most common cause of restricted rooting is physical restriction due to soil compaction, which results in the collapse or diminution of pore spaces and a localised increase in bulk density. Once pores have been compacted to less than about 0.2-0.3 mm diameter, it is difficult for crop roots to freely penetrate the soil. Although the strength of compacted layers decreases as soil water content increases, a high water content can quickly limit the supply of oxygen to roots, so that roots then become restricted by a lack of oxygen.

Certain crops, such as cotton and sunflower, appear to be more susceptible to restricted rooting from compacted layers than others. Compaction often reduces pore sizes sufficiently to inhibit root penetration but not sufficiently to affect the drainage of water through the soil. Pores of 0.2-0.3 mm diameter can restrict roots but water can drain under gravity through pores as small as 0.01 mm diameter. In mechanised cropping systems the continual use of tillage implements, especially disc ploughs, disc harrows, mould-board ploughs and rotovators, over long periods of time frequently results in the formation of dense plough pans containing few pores large enough to be penetrated by crop roots.

The plough pans develop just below the depth to which the soil is tilled and often have smooth upper surfaces with sealed pores, caused by the smearing action of mould-board ploughs. The degree of compaction depends on the pressure exerted by the implements on the soil.

Land preparation when soils are wetter than the optimum moisture content for tillage promotes soil compaction, because the soils are then much more compressible. This is particularly likely to occur on soils that have deficient drainage, or are difficult to till in a dry state without pulverising because of their very hard consistence (*e.g.* hardsetting soils). Compaction is also more likely when farmers use many passes to prepare the seed bed, or when they have only limited tractor power available and are unable to use wide sets of equipment and therefore produce compacted wheel ruts at closer spacing across the field surface.

Compaction can also develop in the subsoil from the passage of heavy machinery such as combine harvesters and lorries loaded with grain, especially in wet conditions. The degree of compaction will depend on the total axle load of the machinery. Soil compaction can also develop from hand tillage. Thin hoe pans just 2-3 cm thick can develop just below the depth of hoe penetration and thus restrict root penetration. When mounds or ridges are formed every year, the combination of hoeing at the same depth and the traffic of people within the furrows during wet conditions may accentuate the compaction.

A similar effect to that of compaction can occur when structurally unstable soils, known as hardsetting soils, slump on becoming saturated by intense rainstorms to form dense layers. On drying the dense soil layer becomes very hard and restricts root penetration. Restricted rooting may also be caused by naturally occurring dense horizons containing few pores large enough for roots to penetrate. These horizons may be found in soils formed from river, lake or volcanic sediments and in semiarid and arid areas where chemically cemented calcrete and gypsic horizons are formed. In some situations root restriction can be caused by a seasonally fluctuating water table. During the rainy season the crop roots are confined to a shallow zone immediately above the high water table.

If, when the water table falls during the dry season the crop roots have already completed their development, the roots will remain where they were, close to the surface, and without access to available water in the deeper subsoil. Roots may also be restricted to shallow depths by chemical factors, such as the presence of toxic aluminium or manganese, or by severe nutrient deficiencies in the subsoil.

The most obvious indicator of restricted rooting when a crop is present is the distribution of the crop roots.

When roots are physically restricted by a dense layer containing few pores suitable for root penetration, individual roots often develop characteristic growth patterns immediately above the restricting layer as shown in Figure 2. The most common of these is the abrupt change in the direction of growth from vertical to horizontal, and a thickening of roots that do manage to penetrate the restricting layer just above the upper boundary of that layer.

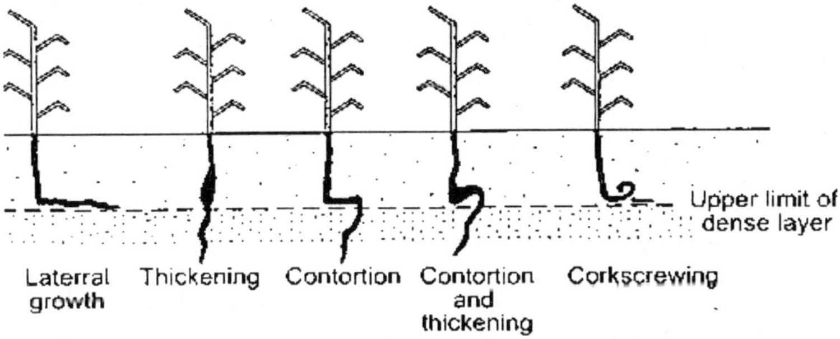

Figure 2. Examples of rooting pattern and growth habit when roots are physically impeded or prevented from penetrating a root-restricting soil layer

In mechanised agriculture, plough pans are usually formed at 12-30 cm depth, depending on the implement used and its normal working depth. Naturally occurring dense layers may occur at any depth. The optimum time to observe roots is after flowering when most of the roots will have largely completed their growth. When no crop is present, it is much more difficult to identify the existence of potentially root-restricting layers in a soil.

However, the rooting pattern of mature weeds, either rooted or uprooted, that remain in the field after the crop has been harvested can be used to reveal the existence of a root-restricting layer. When neither a crop nor weeds are present, the presence of a dense soil layer of high strength and containing very few visible pores will often be a useful indicator. The presence of dense layers is often revealed when digging by the abrupt increase in resistance to the spade or hoe when the restricting layer is reached.

However, sudden increases in soil resistance can also be experienced when the soil changes from moist to dry. To avoid this problem, it is advisable to wet the soil to 30 cm depth two days prior to carrying out the field examination. Physically restricting layers can be identified by the scarcity of visible pores. The smallest pore visible to the naked eye (0.1 mm diameter) coincides reasonably well with the smallest pores into which the seminal roots of cereals (0.1 to 1 mm) and the tap-roots of dicotyledons (0.3 to 10 mm) can penetrate.

When the density of visible pores observed in fragments of the dense layer from a soil pit is less than about six in an area of 10 cm × 10 cm, root restriction is likely to be severe, and responses to breaking up the restricting layer are likely. Other indicators of potentially root-restricting layers that can be used in the field in the absence of a crop are soil strength determined with a penetrometer, and soil bulk density determined from undisturbed soil samples of a known volume. Critical penetrometer resistance and bulk density values at which the roots of most annual crops are restricted have been established for soils of different textures.

9

Soil Management

Good soil condition (tilth) and adequate fertility are basic to sustainable and organic agriculture. One aim of a good soil management programme is to compensate for loss of nutrients through erosion, leaching and immobilisation, and harvested products. High analysis fertilisers compensate for these losses by releasing nutrients to the soil relatively quickly and completely.

In contrast, compensating for these losses with cover crops, mineral powders, and composts, which release nutrients more slowly, requires careful planning and consideration of long-term goals such as improving the structure and biological activity of the soil. Benefits of improved soil accumulate gradually and may become most apparent during dry years.

Managing Physical Characteristics of Soil

Soils are generally classified into four groups: sands, loams, clays, and mucks. Sandy soils dry out and warm up rapidly. Loamy soils contain sand, clay, and silt in nearly equal amounts. While loams are ideal soils for most crops, they are the least common soil in the southern United States. The more common clay soils crust easily. They hold water well and may dry slowly. Clay soils also hold certain nutrients well so that losses to leaching are likely to be lower.

Muck soils are not widespread in the southern United States, but where present, they are well-suited for production of vegetables such as celery and carrots. Like clay soils, muck soils hold water very well. Unlike other southern soils, however, their organic matter content is high, usually greater than 18 percent. Because of their high water-holding capacity and their heat-conducting properties, muck soils warm slowly in the spring.

Compaction

Compaction is a frequent problem in agricultural soils. Crusting, cracks, erosion, subsurface compaction zones (tillage pans), and standing water, especially in wheel tracks, are visible in the field. Plants emerge poorly, grow slowly, vary in size, and have shallow or deformed roots. While compaction of any soil is undesirable, it can be particularly damaging to crop performance in sandy soils.

Surface crusts: A surface crust is a layer of soil more dense than the soil beneath it. Thickness of the crust may vary from 0.5 inch to less than 1/25 of an inch. A dense, impenetrable crust may trap germinating seedlings below the soil surface. On sandy soils with very weak soil structure, crusts usually form as a result of raindrop impact during severe storms, although certain types of overhead irrigation can also contribute to crusting. When the soil dries, dense crusts form because the soil aggregates have broken down. On soils with a tendency to crust, management options include:

- planting seeds at shallow depths,
- protecting the soil surface with mulch or crop residues,

- maintaining a rough soil surface by not over-tilling seed-beds,
- keeping the soil surface moist until seedlings emerge,
- mechanically breaking crusts that form,
- selecting crops such as corn that are able to exert pressure during emergence,
- planting two to four seeds together to increase the pressure they exert during germination, and
- using transplants rather than seeds.

Tillage or pressure pans: Tillage, pressure, or plow pans are compacted layers just below the tillage depth. Pans usually occur when moldboard plows and disk harrows are operated at the same depth over many years. Sandy soils with weak soil structure due to low organic matter content are especially likely to form pressure pans. Pressure pans reduce and, in some cases, even prevent, root penetration so that the plants cannot reach water or nutrients in the soil beneath the compacted zone. Management practices for soils subject to pressure pan formation include reducing or eliminating disk harrowing, using a chisel plow or subsoiler to break tillage pans that have already formed, and limiting vehicle traffic to the smallest possible area of the field.

Surface compaction: Unless soil is very dry or already compacted, both wheel and foot traffic on the soil surface cause some compaction. Even driving once across a field causes compaction. In fact, 80 percent of the compaction resulting from four tractor passes in the same track will be caused by the first pass. The degree of compaction depends on the weight of the load, size of the tire or track, soil moisture content, and soil structure. Surface compaction can be reduced by making fewer trips across the field, pulling a chisel shank behind the wheel on every trip across the field, and restricting all vehicle traffic to the same wheel tracks.

Providing adequate drainage, staying out of wet fields, and increasing soil organic matter content will also help reduce surface compaction. Using flotation tires and keeping tractor wheels out of the furrow may also reduce compaction. A slight underinflation of tires on tractors or implements will also help. In choosing tractors and other farm implements for use where compaction is likely to be a problem, try to minimise the weight pressure applied to any one area and the area to which pressure is applied. Sometimes tradeoffs between these are complex.

Dual tires, for example, reduce pressure, but compact twice as much area. On implements such as large sprayers, using tandem axles following one behind the other instead of dual tires results in a smaller compaction area because the rear tires only add a little compaction. Similarly, a single tire four-wheel drive tractor may cause less compaction than dual tires on a two-wheel drive tractor.

Organic Matter

Adding organic matter to the soil will improve its physical characteristics and increase fertility and biological activity. In general, adding organic matter improves soil structure which enhances soil aeration and promotes penetration of water and roots into the soil. Organic matter bonds sand, silt, and clay particles into soil aggregates, leading to what is called good 'tilth' or the ability of the soil to be worked. Soil water-holding capacity also increases when organic materials are incorporated into the soil. It is very difficult to increase the organic matter of tilled soils, unless legumes or hay crops are included in a rotation or organic material from another area is added to the soil.

However, even if soil organic matter levels are not significantly increased, cycling organic matter still increases biological activity which improves soil chemical and physical properties. For example, the increased microbial activity in worm castings increases soil stability by forming water stable aggregates. Aeration is also increased by worm activity and drainage can be increased four to ten fold.

Fertility Management

Soil Testing

Soil testing is available from state soil testing laboratories and many private testing services. Regular testing not only indicates current soil fertility problems, but also tracks fertility changes over time. In some cases, soil testing saves money by reducing fertiliser use. In the south, most soils that have been in production for long periods are quite high in phosphorus (P) due to years of fertiliser application and low phosphorus mobility in the soil. In 1992, 60 percent of sweetpotato fields, 50 percent of cucumber fields and 70 percent of bell pepper fields in North Carolina needed no additional phosphorus because of the large quantity in the soil. However, undisturbed or non-agricultural soils in the southern United States are frequently low in phosphorus.

Soil test recommendations are usually not cultivar specific. Cultivars vary in their ability to scavenge soil nutrients and some may yield well at less than recommended levels. Soil testing does not provide information about soil structure, nitrogen status, or biological activity, although some estimate of soil organic matter is usually included. The soil's ability to hold nutrients is reflected by soil test measures of cation exchange capacity (CEC). When fields vary naturally in texture and nutrient content it is difficult to get a representative sample for testing.

To overcome this potential problem, guidelines have been developed regarding the depth and number of subsamples collected over a given area. At least 15 to 20 soil cores per sample are usually suggested to obtain a representative sample. Past field history should be considered when interpreting soil test results. This is particularly important when past fertiliser additions have been in the form of organic materials which release nutrients slowly. When such materials are present, soil and laboratory tests may underpredict the amount of soil nutrients actually available to plants over the course of the entire season.

Tissue Testing

Most soil testing services also offer tissue testing. Regular tissue testing is a vital management tool for growers in certain highly controlled environments such as hydroponic vegetable production in greenhouses. In these environments, plants are totally dependent on a constant supply of nutrients and are likely to suffer if they experience even a temporary deficit or imbalance. Outdoors, growers who apply fertiliser through drip irrigation lines also frequently depend on tissue testing to make fertiliser decisions.

The need for tissue testing is less obvious for growers who use slow-release composts, cover crops and manures since fertiliser additions are usually made annually rather than daily or weekly. On the other hand, tissue analysis will help identify toxicity from micronutrients such as copper in poultry litter. Tissue testing will also guide growers in deciding whether to make additional applications of a high-nitrogen manure. In crops such as lettuce and spinach, excess leaf nitrate can cause marketing problems.

Baby food manufacturers, for example, have strict guidelines for nitrate levels present in spinach to be processed. All growers will find tissue sampling valuable if they suspect nutritional problems. For this type of problem, it is important to sample both soil and tissue, since the problem may be nutrient uptake rather than nutrient availability. For example, high salts levels in the soil may damage roots, reducing nutrient uptake directly or by subsequent pathogen infection.

Petiole Sap Testing

Growers who suspect that they have nutritional problems often need an immediate diagnosis. Petiole sap testing kits can be used in such cases for on-farm analysis of nitrate-N and potassium levels. This information would be helpful, for example, to a grower making weekly adjustments in fertiliser injected into drip irrigation systems. When equipment is properly calibrated, results from petiole sap tests are well correlated with laboratory results.

Soil Minerals

Nitrogen: Most organic materials contain nitrogen (N), but the percentage composition varies widely. Average Nitrogen (N), Phosphate (P_2O_5), and Potassiuın (as K_2O) percent composition of various materials. Some of the materials listed, such as apple fruit, contain only trace amounts of N, while others, such as bloodmeal, are relatively high in N. Nutrient sources lists the N percentage and the relative risk of leaching and volatilisation (loss in gaseous form) of selected fertilisers. In interpreting data from the Nutrient sources table, note that the leaching risk for urea and ammonium fertilisers is low only as long as the N remains in the ammonium form. In soils with high biological activity, nitrifying soil microorganisms can rapidly convert ammonium to the highly leachable nitrate form. Nitrogen availability may be reduced, however, by lower biological activity in cold or extremely dry or flooded soils.

Phosphorus: Phosphorus is also available from organic matter additions; P availability increases when conditions are favourable for microbial activity and root growth. In cool weather, lower biological activity results in less P being available than in the same soil under conditions favouring biological activity. Cool soil temperatures also restrict root growth which reduces the amount of nutrients taken up by roots. In the soil, P is found as reaction products which have precipitated out of solution and as held or adsorbed forms on surfaces of various soil minerals.

Soil testing before adding P is particularly important because phosphorus levels of southern coastal plain soils that have been farmed for many years may not need additional phosphorus for several years. On the other hand, southern soils which have not been fertilised are often quite low in P. Since phosphorus has only limited mobility in the soil, incorporation is a better application method than side dressing or broadcasting. Sidedressing P may concentrate rooting in the area fertilised, thereby increasing the need for irrigation because of restricted root growth outside the area of P application.

Potassium: In the soil, potassium is held by adsorption on soil particles, cation (charged particle) exchange, and inclusion in other reserve pools. When root development is good, plants are better able to exploit potassium reserves in the soil. Organic material is generally a good source of potassium. Although granite dust and greensand contain only a small percent of immediately available potassium and some trace elements, they release potassium slowly through mineralisation. These materials are most beneficial when used to build reserves in the soil.

Lime and other calcium sources: Agricultural lime, chemically described as the oxide, hydroxide, or carbonate of calcium and/or magnesium, is used to correct soil pH. The best forms of lime to use depend on the situation. Both dolomitic and calcitic limestone have approximately the same liming capability, but dolomitic limestone has a higher magnesium content and is useful where soil magnesium levels are low. The amount of lime to apply should be determined by a soil test. Although agricultural lime is the most commonly used method of raising soil pH, finely ground shells, wood ash, and other materials will also raise soil pH levels.

Gypsum (calcium sulphate) supplies calcium, but has no effect on soil pH. Gypsum can reduce erosion on heavy clay soils by improving soil structure and increasing water infiltration. To reduce erosion and soil crusting, gypsum is spread over the soil rather than incorporated. Fine particle size gypsum is most effective. Because gypsum is more soluble than limestone, the calcium in gypsum leaches through the soil and increases calcium availability at deeper levels in the soil. Adding calcium in the subsoil can help reduce aluminum toxicity. This will allow root growth at deeper levels, which improves drought tolerance.

There may be little or no effect the first season, but beneficial effects of gypsum on yield in calcium-deficient soils may persist for six years or more after application. However, calcium interactions with other cations should also be considered. Calcium from gypsum may replace magnesium and potassium in the topsoil, which makes deep, sandy soils, which have a low Cation exchange capacity, vulnerable to loss of magnesium and potassium by leaching at high calcium levels. Because of these interactions, timing of lime application in relation to other fertiliser applications can be important.

Secondary nutrients and micronutrients: Secondary nutrients (calcium, sulphur, and magnesium) and micro-nutrients (boron, copper, iron, manganese, molybdenum and zinc and chlorine) are essential for growth, but required in smaller quantities than N, P, and K. Additions of micronutrients should be made only when a deficiency is indicated by a soil test or plant tissue analysis. Many factors determine whether organic matter increases or decreases the availability of micronutrients.

Metallic cations, including iron, manganese, zinc and copper, can be bound to chelates (organic compounds that are able to bind tightly with cations). In this form they are partially protected from changes in pH and from forming precipitates. Iron, for example, is typically added to fertiliser solutions in chelated form. However not all organic compounds that chelate micronutrients make them more available. Micronutrients may be bound so tightly to organic matter in organic soils (greater than 18 percent organic matter), that crops suffer micronutrient deficiencies.

Copper is one micronutrient that is often unavailable and deficient in organic soils. With some nutrients, however, lack of availability is an advantage because it reduces potentially toxic levels of micronutrients. Manure additions, for example, rarely lead to toxic levels of micronutrients even though fairly high levels of micronutrients are present in manure.

Organic Matter Additions

In general, organic matter additions to a soil will increase its ability to hold nutrients in an available state. Organic matter additions will also increase soil biological activity which will affect the availability of nutrients in the soil. Soil which has received organic matter has increased microbial populations and more varied fungal species than soils receiving synthetic fertiliser applications. The long-term objective of organic matter addition is to build up soil humus. Humus is the more or less stable fraction of the soil organic matter remaining after decomposition of plant and animal residues.

Manure

Many soils in the southern United States have poor tilth. Manures and other materials with a high organic component are especially useful where topsoil has been lost. They are also beneficial in soils with micronutrient deficiencies or where specific nutrient needs cannot be determined by soil testing. Manures and composted materials also are used to improve the water-holding capacity of sandy soils and to improve the structural stability and tilth of heavy clay soils. Manure, along with crop residues, is a primary means of replenishing soil organic matter.

Although crops respond readily to the nutrients in manure, the total yield benefits from manure sometimes take three or more years to become apparent. Although a portion of the nutrients and organic matter in manure is broken down and released during the first year or two, a portion is also held in humus-like compounds which decompose more slowly. Uncomposted manures are difficult to apply, not only because of their bulk, but because it is easy to apply more nitrogen than the plants can absorb. For this reason direct application of manure or other raw animal wastes is not recommended nor is it usually allowed in certified organic production.

Direct application can lead to problems of excess nitrates in the plant and runoff of nitrates into surrounding water supplies. Excessive raw manure can burn plants and lead to toxic levels of nitrates in leafy greens. Nitrogen, the most unstable nutrient in manure, can easily be lost by volatilisation in the form of ammonia gas. For example, when manure is spread on frozen ground, 25 percent of the nitrogen can be lost immediately and 75 to 80 percent can be lost within two weeks. Delaying application until manure can be immediately incorporated into the soil will lessen the amount of nitrogen lost. Determining the correct amount of manure to apply is difficult.

Manure samples should be analysed for nutrient content and levels of metals such as copper which are often present in poultry litter. Nitrogen available to the plant is lower than the content in the samples since some loss occurs through volatilisation with spreading and since only a portion of the organic N becomes available to the plants through mineralisation during the growing season.

Also, the rate of manure application needed to supply the nitrogen needs of the crop will usually supply phosphorus and potassium in amounts in excess of those the plant can use. This excess application generally does not affect crop growth but can, in the case of phosphorus, pollute water if runoff or erosion occurs. Phosphorus runoff can be minimised by controlling erosion with cover crops or mulches. Average nutrients available in manure from different sources is listed here.

Composts

In comparison to chemical fertilisers, composts have relatively low amounts of nitrogen, phosphorus, and potassium. However, the total amount available over time is greater than that immediately available. Some nutrients are held in humus-like compounds which decompose very slowly so that their effects on soil continue for years after application. Composts can be made on-farm, but they are also available from municipal and commercial sources.

Municipal sources usually consist of yard wastes, minimally processed, and distributed at low or no cost. These composts vary in quality, but tend to have low fertility and such high C:N ratios that little N is available. Since plastic, trash, and industrial waste may also turn up in municipal wastes, some state organic certification programmes do not allow their use. Commercially composted chicken and turkey litter is also increasingly available in the south. Poultry litter usually has a higher nutrient level than other composts and few weed seeds, but may contain feed additives, including heavy metals such as copper. Composting raw materials before applying them to the soil has several advantages. It:

- prevents the loss of N through ammonia gas (NH_3) by fixing N into organic forms.
- allows the safe use of materials such as cannery wastes and sawdust that otherwise tie up soil nitrogen.

- distributes mineral soil amendments more evenly than direct application.
- reduces pathogen populations when temperatures reach 150 degrees F and destroys weed seeds when temperatures reach 175 degrees F.
- reduces the volume of material that has to be applied.
- stores and handles more easily.
- reduces N availability below levels that will burn plants.

Composting also has several disadvantages. It:

- is not as effective as raw organic matter in improving soil structure. As microorganisms work to decompose raw organic matter, they excrete gels and slimes that bind soil particles together and enhance soil structure.
- results in loss of nitrogen as NH_3 when the compost is turned.
- reduces nitrogen availability in comparison to the raw material from which it was made.

Successful composting requires a moisture content of 50 to 60 percent, oxygen content of 5 percent or more, and a carbon to nitrogen ratio of 30:1. If the oxygen content falls below 5 percent, decomposition becomes anaerobic, slows down, and creates unpleasant smells. The optimum temperature is from 122 to 140 degrees F. Under optimal conditions, the temperature of the compost increases rapidly as heat is given off by the composting process. Within a few days it will reach its maximum temperature. At this point, organic matter burns up rapidly and the pile is said to 'cook.'

Within a week or so the pile starts to cool. The temperature during the heating part of the cycle is critical. If temperatures are hot enough, pathogens (150 degrees F required) and weed seeds (175 degrees F required) are killed. However, if temperatures are too high, the process self-destructs, killing the various bacteria, fungi, and actinomycetes whose activities drive the composting process. Thus the pile is turned or mixed to maintain temperatures from 122 to 140 degrees F and to keep oxygen above 5 percent.

As the pile cools, decomposition continues at a slower rate. This maturing stage may take weeks or months. At some point, however, the compost temperature returns to that of the surrounding air and soil and further decomposition is minimal. This stable compost will not heat further when the pile is turned. Usually the stable product has a high C: N ratio (greater than 30:1), but there is no standard value for this ratio. Many materials can be used to produce composts. Examples of the percentage of N, P, and K in various raw materials. Even compounds like pond mud obtained from cleaning out farm ponds can be used as a soil amendment. If not composted, however, this mud will have the same effect as raw manure.

Organic Mulches

A mulch is any material placed on the soil surface to protect the soil from the adverse effects of raindrops, freezing, and water loss. Mulches are also used to control weeds and reduce erosion. Organic mulching materials will break down over time, contributing organic matter to the soil.

Management of Biological Characteristics

A complex array of soil-dwelling plants and animals

decomposes organic matter, mineralises organic forms of nutrients, and fixes nitrogen. These beneficial biological activities enhance the soil's ability to release nutrients needed for plant growth and to break down plant residues. Tillage practices that aerate the soil enhance biological activity by providing oxygen and mixing organic matter throughout the tilled area. On the other hand, broad-spectrum fumigants such as methyl bromide kill beneficial soil organisms along with soil pests.

Both ammonium and nitrate are readily available sources of nitrogen to plants and soil microorganisms. A variety of soil microorganisms convert organic nitrogen to ammonium, but only specific (nitrifying) bacteria convert ammonium to nitrate (nitrification). Nitrifying bacteria are most effective in well aerated soils at soil temperatures of 80 to 90 degrees F, when the C:N ratio is low to medium and when soil moisture is adequate for plant growth. High activity of these bacteria increases N availability, but the nitrates released by nitrifying bacteria are highly soluble and subject to leaching.

Thus, if not absorbed by crop or covers, nitrate may be lost from the field. In addition, other soil bacteria (denitrifiers) reduce nitrates to elemental nitrogen or nitrous oxide which are lost into the atmosphere as they are volatilised. Denitrification is most likely to occur in poorly drained soils where oxygen levels are low (anaerobic conditions). Not only do naturally occurring microorganisms play an essential role in the nitrogen cycle, they have also been reported to decrease some populations of pathogenic bacteria and fungi.

For example, the fungus Gliocladium virens, which tends to increase with increased levels of organic matter, controls damping-off pathogens. G. virens is also available commercially. Agrobacterium, a naturally occurring soil microorganism, reportedly restricts growth of Fusarium, a fungal pathogen causing a number of diseases in vegetable crops. Beneficial microorganisms in the root zone, particularly species of Pseudomonas, have been associated with a decrease in take-all decline of wheat. Over time, two species of soil-dwelling amoebae in Canada reduced populations of two root pathogens.

Earthworms break up organic matter, combine it and humus with soil particles, and enhance microbial activity. Soils with high earthworm populations or soils containing earthworm castings generally have a greater ability to hold water and plant nutrients. Cation-exchange capacity, exchangeable calcium, magnesium, and potassium, and available phosphorus are generally higher in soils with earthworms. Earthworm activity has also been associated with reduced severity of several diseases.

In Australia, Rhizoctonia solani caused less damage to wheat seedlings when earthworms were present. Mechanical soil disturbance also reduces disease severity, and it is possible that the physical mixing of soil by the worms reduces pathogen severity. The presence of earthworms also enhanced the movement and colonisation of wheat roots by the biological disease control microorganism Pseudomonas corrugata. When earthworms are introduced to the soil, they spread rapidly from the inoculation point—up to 36 feet per year. In most soils, however, if organic matter increases, earthworm populations will increase rapidly. Continuous cropping without adding organic matter usually reduces earthworm populations to a very low level, as do copper fungicides, soil fumigation, and frequent tillage.

Management Steps to Improve Soil Quality

Assess Soil Health and Biological Activity

A basic soil audit is the first and sometimes the only

monitoring tool used to assess changes in the soil. Unfortunately, the standard soil test done to determine nutrient levels (P, K, Ca, Mg, etc.) provides no information on soil biology and physical properties. Yet most of the farmer-recognised criteria for healthy soils include, or are created by, soil organisms and soil physical properties.

Direct assessment

Some quick ways to identify a healthy soil include feeling it and smelling it. Grab a handful and take a whiff. Does it have an earthy smell? Is it a loose, crumbly soil with some earthworms present? Look at the surface and see if it is crusted, which tells something about tillage practices used, organic matter, and structure. Push a soil probe down to 12 inches, lift out some soil and feel its texture. If a plow pan were present it would have been felt with the probe. Turn over a shovelful of soil to look for earthworms and smell for actinomycetes, which are microorganisms that help compost and stabilise decaying organic matter. Their activity leaves a fresh earthy smell in the soil.

Two other easy observations are to count the number of soil organisms in a square foot of surface crop residue and to pour a pint of water on the soil and record the time it takes to sink in. Comparisons can be made using these simple observations, along with Ray Weil's evaluation above, to determine how farm practices affect soil quality.

A simple erosion demonstration

This simple procedure demonstrates the value of ground cover. Tape a white piece of paper near the end of a three-foot-long stick. Hold the stick in one hand so as to have the paper end within one inch of a bare soil surface. Now pour a pint of water onto the bare soil within two to three inches of the white paper and observe the soil accumulation on the white paper. Tape another piece of white paper to the stick and repeat the operation, this time over soil with 100% ground cover, and observe the accumulation of soil on the paper. Compare the two pieces of paper. This simple test shows how effective ground cover can be at preventing soil particles from detaching from the soil surface.

Tools for build soil

Can a cover crop be worked into your rotation? How about a high-residue crop or perennial sod? Are there economical sources of organic materials or manure in your area? Are there ways to reduce tillage and nitrogen fertiliser? Where feasible, bulky organic amendments may be added to supply both organic matter and plant nutrients. It is particularly useful to account for nutrients when organic fertilisers and amendments are used. Start with a soil test and a nutrient analysis of the material you are applying.

Knowing the levels of nutrients needed by the crop guides the amount of amendments applied and can lead to significant reductions in fertiliser cost. The nutrient composition of organic materials can vary, which is all the more reason to determine the amount you have with appropriate testing. In addition to containing the major plant nutrients, organic fertilisers can supply many essential micronutrients. Proper calibration of the spreading equipment is important to ensure accurate application rates.

Cover crops and green manures

Many types of plants can be grown as cover crops. Some

of the more common ones include rye, buckwheat, hairy vetch, crimson clover, subterranean clover, red clover, sweet clover, cowpeas, millet, and forage sorghums. Each of these plants has some advantages over the others and differs in its area of adaptability. Cover crops can maintain or increase soil organic matter if they are allowed to grow long enough to produce high herbage. All too often, people get in a hurry and take out a good cover crop just a week or two before it has reached its full potential.

Hairy vetch or crimson clover can yield up to 2.5 tons per acre if allowed to go to 25% bloom stage. A mixture of rye and hairy vetch can produce even more. In addition to organic matter benefits, legume cover crops provide considerable nitrogen for crops that follow them. Consequently, the nitrogen rate can be reduced following a productive legume cover crop taken out at the correct time. For example, corn grown following two tons of hairy vetch should produce high yields of grain with only half of the normal nitrogen application. When small grains such as rye are used as cover crops and allowed to reach the flowering stage, additional nitrogen may be required to help offset the nitrogen tie-up caused by the high carbon addition of the rye residue.

The same would be true of any high-carbon amendment, such as sawdust or wheat straw. Cover crops also suppress weeds, help break pest cycles, and through their pollen and nectar provide food sources for beneficial insects and honeybees. They can also cycle other soil nutrients, making them available to subsequent crops as the green manure decomposes.

Humates

Humates and humic acid derivatives are a diverse family

of products, generally obtained from various forms of oxidised coal. Coal-derived humus is essentially the same as humus extracts from soil, but there has been a reluctance in some circles to accept it as a worthwhile soil additive. In part, this stems from a belief that only humus derived from recently decayed organic matter is beneficial. It is also true that the production and recycling of organic matter in the soil cannot be replaced by coal-derived humus. However, while sugars, gums, waxes and similar materials derived from fresh organic-matter decay play a vital role in both soil microbiology and structure, they are not humus.

Only a small portion of the organic matter added to the soil will ever be converted to humus. Most will return to the atmosphere as carbon dioxide as it decays. Some studies have shown positive effects of humates, while other studies have shown no such effects. Generally, the consensus is that they work well in soils with low organic matter. In small amounts they do not produce positive results on soils already high in organic matter; at high rates they may tie up soil nutrients.

There are many humate products on the market. They are not all the same. Humate products should be evaluated in a small test plot for cost effectiveness before using them on a large scale. Salespeople sometimes make exaggerated claims for their products. ATTRA can provide more information on humates upon request.

Minimise nitrogen use

If at all possible, add carbon with nitrogen sources. Animal manure is a good way to add both carbon and nitrogen. Growing legumes as a green manure or rotation crop is another way. When using nitrogen fertiliser, try to do it at a time when a heavy crop residue is going onto the soil, too. For example, a rotation of corn, beans, and wheat would do well with nitrogen added after the corn residue was rolled down or lightly tilled in. Spring-planted soybeans would require no nitrogen.

A small amount of nitrogen could be applied in the fall for the wheat. Following the wheat crop, a legume winter-annual cover crop could be planted. In the spring, when the cover crop is taken out, nitrogen rates for the corn would be reduced to account for the nitrogen in the legume. Avoid continual hay crops accompanied by high nitrogen fertilisation. The continual removal of hay accompanied by high nitrogen speeds the decomposition of soil organic matter. Heavy fertilisation of silage crops, where all the crop residue is removed (especially when accompanied by tillage), speeds soil decline and organic matter depletion.

Bibliography

Allen M. F. et al., "Soil Microorganisms", In L. Walker (ed.) Ecosystems of disturbed ground. Elsevier Press, 1998.

Bardgett, R.D. and Wardle, D.A., "Herbivore mediated linkages between aboveground and belowground communities", *Ecology,* 2003.

______________ ., "Causes and consequences of biological diversity in soil", *Zoology,* 2002.

______________ ., Streeter, T. and Bol, R., "Soil microbes compete effectively with plants for organic nitrogen inputs to temperate grasslands", *Ecology,* 2003.

Clark, M.S., W.R. Horwarth, C. Shennan, K.M. Scow, "Changes in soil chemical properties resulting from organic and low-input farming practices", *Agronomy Journal,* 1998.

Cole, L.C. , Staddon, P.L. Sleep, D. and Bardgett, R.D., "Soil animals influence microbial abundance, but not plant-microbial competition for soil organic", *Functional Ecology,* 2004.

Hu, S., N. Grunwald, A.H.C. van Bruggen, G.R. Gamble, L.E. Drinkwater, C. Shennan, and M.W. Demment. "Short-term effects of cover crop incorporation on soil carbon pools and nitrogen availability", *Soil Science Society of America Journal,* 1997.

Mitchell, J.P., C.D. Thomsen, W.L. Graves and C. Shennan, "Cover crops for saline soils", *Journal of Agronomy and Crop Science,* 1998.

Morris, S. J., T. Zink, K. Conners, and M. F. Allen, "Comparison between fluorescein diacetate and differential fluorescent staining procedures for determining fungal biomass in soils", *Applied Soil Ecology,* 1997.

Scow, K. M. and M.R. Werner, "The soil ecology of cover cropped vineyards", In C. Ingels et al. (eds.). *Cover Cropping in California Vineyards: A Grower's Handbook,* 1998.

Smith, R.S, Shiel, R.S., Bardgett, R.D., Millward, D., Corkhill, P., Rolph, G., Hobbs, P.J., "Diversification management of meadow grassland: plant species diversity and functional traits associated with change in meadow vegetation and soil microbial communities", *Journal of Applied Ecology,* 2003.

Wardle, D. A. and Bardgett, R. D., "Indirect effects of invertebrate herbivory on the decomposer subsystem", In: *Insects and Ecosystem Functioning* (Ed. by W. Weisser and E. Siemann) *Ecological Studies,* Springer, Heidelberg, in press, 2004.

Werner, Matthew R., "Soil quality characteristics during conversion to organic orchard management", *Applied Soil Ecology,* 1997.

Index